Peter Jackson-Lee

HISTORY OF THE MERSEY TUNNELS

AUSTIN MACAULEY PUBLISHERS™
LONDON * CAMBRIDGE * NEW YORK * SHARJAH

Copyright © Peter Jackson-Lee (2018)

The right of Peter Jackson-Lee to be identified as author of this work has been asserted by him in accordance with section 77 and 78 of the Copyright, Designs and Patents Act 1988.

All rights reserved. No part of this publication may be reproduced, stored in a retrieval system, or transmitted in any form or by any means, electronic, mechanical, photocopying, recording, or otherwise, without the prior permission of the publishers.

Any person who commits any unauthorised act in relation to this publication may be liable to criminal prosecution and civil claims for damages.

A CIP catalogue record for this title is available from the British Library.

ISBN 9781788781152 (Paperback)
ISBN 9781788781169 (Hardback)
ISBN 9781788781176 (E-Book)

www.austinmacauley.com

First Published (2018)
Austin Macauley Publishers Ltd
25 Canada Square
Canary Wharf
London
E14 5LQ

Over the last three years, I have met and corresponded with numerous people, most of whom were only too glad to pass on their information and experiences, without whom, the immense amount of details in the book would have been impossible to attain. I do apologise if I have forgotten anyone, and I would like to thank the following for their time and assistance. Carl Lecky MBE, who is an accomplished author himself and gave me the inspiration to write the book; Sheena and the staff at the Birkenhead Reference Library; Will, Julie and Fran at Wirral Archives; and the various staff at the Liverpool Museum all of whom I would like to thank for their patience and understanding for my numerous requests for records and assistance; Jeremy Wilkinson – Mines and Quarries of North Wales.

Not forgetting Elaine, my dearest wife, for putting up with me over the last three years whilst I researched and wrote this book in whatever spare time I had between working and other everyday commitments. I was frequently missing for large period of times only to be found typing or off on further research trips.

Lastly, and by no means least, I would like to dedicate this book to the men who have died during the construction of both tunnels. Our heartfelt thanks should go to these men in producing such marvels of engineering and making the life of people on both sides of the river easier.

Table of Contents

Prologue   7

Introduction   8

Chapter 1   15
History of Tunnel Construction

QUEENSWAY TUNNEL   21

Chapter 2   23
Queensway Tunnel – The Beginning

Chapter 3   27
The Tunnel Makers

Chapter 4   32
Agreement of the Various Parties

Chapter 5   44
Finances

Chapter 6   63
Purchase of Property

Chapter 7   69
Full Scale Experiment

Chapter 8   72
Start Tunnelling

Chapter 9   112
Drainage Services and Ventilation

Chapter 10   **139**
    Finishes

Chapter 11   **157**
    Queensway Tunnel Opening Ceremony

WALLASEY TUNNEL   **177**

Chapter 12   **179**
    Kingsway Tunnel

Chapter 13   **235**
    Bidston Moss Piling

Chapter 14   **242**
    Kingsway Tunnel Ventilation

Chapter 15   **247**
    Mersey Tunnel Police

Chapter 16   **252**
    Tolls Finances Maintenance and Repairs

Chapter 17   **301**
    Second World War

Chapter 18   **313**
    Queensway Tunnel Approach Scheme, July 1969

Chapter 19   **331**
    Emergency Exits

Chapter 20   **342**
    Tunnellers Memorials

Chapter 21   **348**
    Update on Companies

Bibliography   **360**

Prologue

I have travelled (as have many people in Liverpool and Wirral) through the Mersey Tunnels on a daily basis for many years and have always been fascinated by their construction. This marvel of engineering has given birth to the modern tunnels we now take for granted. No TBMs (Tunnel Boring Machines) and mechanically placed large concrete lining sections were used in the original Queensway Tunnel, it was all by hand and honest hard graft. That is not to say today's tunnellers have it easy, because they don't. They have the same problems with ground faults, working at great depths, and the horror of fires and tunnel collapsing around them, not to mention decompression sickness, which is normally associated with divers. They are just able to undertake the job quicker and more efficiently, and with larger diameter tunnels which can be cut within centimetres of an existing structure's foundations. In some cases, a tunnel can be cut in areas that they would never have dreamt of during the construction of the Queensway Tunnel.

Throughout this book, you will see figures and measurements in brackets next to the original figure. These brackets have brought the information up to date from feet and inches to metres and the costs of the tunnels in 1934 and 1971/2 to the present day. Some of the figures from the original Queensway build costs would be seen as eye-watering today. That is not to say they were not seen as such in the early days of tunnelling but it just brings the mind to focus on the eye watering cost of the HS2 (£188 Million) and London's Crossrail underground extension (The current cost estimate for the project is £12 billion in 2012 prices. This rises to £20 billion).

Imagine if, one day you raised from your weary sleep to find the tunnels were no longer there or in use. You would have to try and use the ferries to get to work, college, university, by trying to gain access to the ferry terminals at Woodside and Seacombe, let along finding somewhere to park. Both sides of the river would be completely different as most of Wirral travels to Liverpool in the morning and back in the evening, the mid Wirral motorway (M53) would have no access to Liverpool or the docks, let along the rest of the world from both sides of the river. The old Runcorn Bridge, Silver Jubilee Bridge, along with the new tolled Mersey Gateway Bridge would be exceptionally busy and effectively car parks in rush hour. This would be a glimpse at how the two holes in the ground under the River Mersey would make life difficult if they were not there and how they are a major asset to the people of Merseyside and the surrounding area.

Introduction

For those of you who are unfamiliar with the Mersey Tunnels or their geographical location, you may know the area better for the Beatles or the famous outline of the Pier Head, known locally as the three graces, Cunard Building, Royal Liver Building and Port of Liverpool Building.

As you travel along the river, heading out to the Irish Sea, you will have Liverpool and its three graces on the right, with the Wirral Peninsula (Wirral) to the left. For many years, the only way to traverse the river and reach either the Wirral or Liverpool sides was to travel by ferry. By the early 1900s, the route and method of transportation for goods, which included horse drawn carts, and passengers was fast becoming a congested route.

After the River Dee silted up and Chester as a usable port became impossible, the fortunes for Liverpool increased and blossomed into the international port and travel destination we see today. A crossing for the River Mersey was first discussed around 1825 by Mark Brunel whilst he was undertaking the Wapping – Rotherhide Tunnel. There are now four ways to traverse the river, Ferry, Railway Tunnel (Opened in January 1886 by the Prince of Wales (the future King Edward VII) as the first underwater railway and the first to be converted from steam to electricity) and the two Mersey Tunnels (Queensway in Birkenhead and Kingsway in Wallasey). There is another way to cross the river but that would see a lengthy trip to Runcorn, which in the turn of the century would take the best part of the day to get there and back. Even today, the Runcorn (Silver Jubilee) Bridge is undergoing a full refurbishment following the opening of the Mersey Gateway Bridge to prevent it becoming a congested bottleneck in the future. If the Queensway (first tunnel) was being built today, a 44 ft (13m) diameter tunnel under a river estuary, with two branches, underwater junctions and four tunnel portals to be squeezed into city centre locations would be an incredibly large-scale and possibly excessively expensive project.

Unlike any tunnel projects being undertaken today, the only viable prospect in 1923 was to excavate by hand. No tunnel of comparable diameter had been built before, and nothing existed to match its length or its complexity. It is for this reason it was known at the time as the eighth wonder of the world. No mechanism existed by which central government could plan such large-scale schemes. The railways, which held enormous political sway, were vociferous opponents, given that they operated the only passenger sub surface existing river crossing. The odds were, really, stacked against its construction.

All the same, the Mersey Tunnel Joint Committee was formed in 1923 and within a little more than two years' feasibility and design work, they started progress on the tunnel. When it opened in 1934, it was the engineering marvel of the world

The story of both the Mersey's Queensway and Kingsway Road Tunnels is a bit of a long story of progression through the ages as the areas grew. Ever since the Benedictine Monks at the Birkenhead Priory close to what is now Cammell Lairds Ship Repairs started to row people across the river in 1150, there has always been a wish to cross the river in one way or another. The Monks started rowing people across the river for a small fee, which at times on a tidal river was very dangerous.

At this time, the Mersey was considerably wider with sand dunes and marshes to the north leading up to Ainsdale beach and sandstone cliffs and shorelines to the south near Otterspool. The only suitable landing point for the ferry was in the Pool, near the site of the present Merseyside Police Headquarters. Weather often stopped crossings and passengers were delayed for days, taking shelter at the priory.

In 1317, a royal licence was issued, granting permission to the Priory to build lodging houses for men crossing the river at Woodside. King Edward II visited Liverpool in 1323, and the royal accounts show that he used local ferrymen to sail up the river to Ince. In 1330, his son Edward III granted a charter to the Priory and its successors forever: 'the right of ferry there… for men, horses and goods, with leave to charge reasonable tolls'.

At the time, there was only a small hamlet at Birkenhead, and a slightly larger village at Liverpool. The monks of Birkenhead Priory operated a ferry service until the dissolution of the monasteries and the priory's destruction by Henry VIII's troops in 1536. Ownership reverted to the Crown. By the 18th century, the commercial expansion of Liverpool and the increase in stagecoach traffic from Chester spurred the growth of the transportation of passengers and goods across the river. Ferry services from Rock House on the Wirral were first recorded in 1709.

By 1753, the Wirral side of the Mersey had at least five ferry houses at Ince, Eastham, the Rock, Woodside and Seacombe. The service from New Ferry to Liverpool was first mentioned in 1774.

Due to the natural progression and the growing population, not to mention businesses in and around the docks, things had progressed by 1886 when the railway opened and directed people from Wirral to Liverpool via an under-river tunnel. This took a small amount of traffic away from the ferries, but they still had all the carts, horses and other forms of business traffic across the river.

Then in 1934, that all changed again when the Queensway Tunnel at Birkenhead opened (it had been in use since 17 December 1933) followed by the Wallasey Kingsway in 1971. This is the story of how both tunnels have come about and allowed the areas of Wirral and Liverpool to grow.

The term 'left-footers' was applied to tunnel workers because of the noise they made as they walked to and from work across the city's cobbled streets. Digging the tunnel involved using the left foot (for most workers) driving their shovels and spades into the ground. The result was workers' left shoes were constantly in need of repair or replacement. To slow the wear and tear the men attached pieces of iron onto the soles of their shoes. Consequently, this caused the distinctive sound as they clattered through the streets and their left-footer nickname.

Work started on the tunnels in December 16th 1925. Princess Mary set the pneumatic drills in motion to enable the first shaft to be dug on the surviving portion of the Old Georges Dock at the Pier head. It was from this side of the river (Liverpool) that two pilot tunnels, one above the other were excavated below the River Mersey. Similar excavations were happening on the Birkenhead side at Morpeth Dock. Excavation was made using a combination of drilling and explosives with a maximum of 1,700 men working on the tunnel at one time. Seventeen of these men lost their lives and they are commemorated later in this book, as are those who lost their lives in the Wallasey Tunnel.

The following images were seen in a Birkenhead Advertiser Supplement June 14th 1934 for the opening of the Queensway Tunnel:

An "Advertiser" cartoon which helped to stimulate interest in the preliminary talks which made the Tunnel possible.

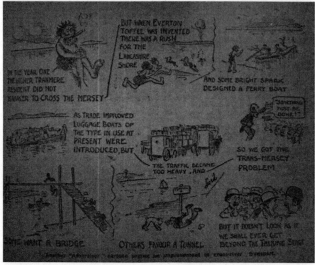

Another "Advertiser" cartoon urging an improvement in cross-river transport.

What might have been! A Mersey Bridge design in 1921.

Mersey Tunnels have a 'Tunnel Tour' and this has been described as a 'Hidden Gem' by Visit England Visitor Attraction Quality Scheme in awards in 2015. The tour offers people the chance to view the tunnels from the inside out and you will see the inner workings of the tunnels and get to visit the areas below the roadway and ventilation systems. The accolade is over and above the Visit England's Visitor Attraction Quality Assurance Service accreditation (VAQAS) that Mersey Tunnels have been associated with.

The Mersey Tunnels see up to 90,000 vehicles a day cross the River Mersey via the Queensway (Birkenhead) and Kingsway (Wallasey) tunnels. Queensway was the original tunnel opened in 1934 by King George V and the Kingsway was opened in 1971 by Queen Elizabeth II. The Mersey Tunnels are regarded as the safest tunnels in the UK and the Queensway, the safest tunnel in Europe despite its 80+ years of age.

Historic England (Formerly English Heritage) Statements have the various tunnel elements noted as follows

Queensway Tunnel, Birkenhead
Pacific Road Ventilation Station

Ventilation station for the Mersey Road tunnel. 1925-34. Sir Basil Mott and J.A. Brodie, engineers, Herbert Rowse, architect. Steel framed with brick cladding. Composed of a series of geometric blocks grouped around the giant main tower. Decoration provided by nogged projecting brick courses on main tower in the form of a cross, the tops of each block stressed with bands of stepped and nogged brick courses. Brick bands give a rusticated effect to base. Doors with chevron decoration in tower base. Houses giant fans used in ventilation, the largest of a series of three towers on the Birkenhead side of the tunnel.

Sidney Street Ventilation Station

GV II Ventilation Station to the Mersey Road Tunnel. 1925-1934. By Sir Basil Mott and J.A. Brodie, as engineers and Herbert J. Rowse, architect. Brick faced, steel framed construction. Massive twin towers, largely blind but with double doors in south elevation divided by three blind lancet slits. Doors have chevron bands, and architraves formed of brick in rippled courses, with central wing motifs. Ribbed brick quoins to towers, blind lancets and geometrical ribbed decoration towards the top. Built to house fans as part of the ventilation system for the road tunnel.

Taylor Street Ventilation Station

Ventilation Station of the Mersey Road Tunnel. 1925-34. By Sir Basil Mott and J.A. Brodie, engineers, and Herbert J. Rowse, architect. Brick faced steel framed construction. Tower comprised of a series of grouped geometric blocks, largely blind but with small doorway in eastern elevation, and double doors in northern elevation. Doors themselves original features, with bands of chevron and scallop decoration. Architraves formed from rippled coursing of brick, with wing motifs in centre. Some ornamentation in the brickwork, with diamond frieze at base and giant cross motif in the largest of the blocks. Chevron bands of brickwork mark the tops of the towers, and the largest block is marked by a frieze of recessed panels or slits. Built to house fans as part of the ventilation system for the road tunnel. (The Buildings of England: Pevsner N and Hubbard E: Cheshire: Harmondsworth: 1971).

Former Dock Entrance, Rendel Street

Dock entrance to Mersey Road Tunnel, now disused. 1931-34. By Sir Basil Mott and J.A. Brodie as engineers, Herbert J. Rowse, architect. White Portland stone. Retaining walls marking approach each side of tunnel portal, with low terminal blocks at each end, and tunnel entrance. Modern style with Egyptian detailing. Tunnel entrance with low relief shield flanked by winged beasts over. Scalloped cornice to retaining walls built on curve to each side. Terminal blocks stepped in plan and battered in section. Narrow bands of window, and stone grid-work with ribbed cornice band with arrow motif in low relief.

Entrance to Mersey Tunnel, King's Square

Tunnel entrance. 1925-34. Engineered by Sir Basil Mott and J. A. Brodie, with Herbert J. Rowse as architect. Faced in white Portland stone. Egyptian style. Two flanking lodge towers and walls leading downwards towards portals. Lodge towers have round-headed arched entrances flanked by fluted engaged shafts, and wing motifs in low relief over. Scalloped frieze and high blocking course above. Flanking walls link to retaining walls of tunnel entrance, with chevron frieze. Wing motif over portals. Original height of westernmost lodge tower reduced during alterations, which have included additional building against the flanking wall and partially glazed infilling of formerly open ground floor of towers.

Monument to the Building of the Mersey Tunnel, Chester Street

Monument commemorating construction of Queensway (Mersey) Tunnel. 1934 by Herbert Rowse. White ashlar base with polished black granite shaft. Tall fluted shaft on high base with chevron and acanthus decoration in bands at base and top. It is capped by a fluted glass bowl out of which a ribbed cap carries a banded sphere. Base is inscribed with names of the engineer and architect of the tunnel, Sir Basil Mott J.A. Brodie, engineers, Herbert Rowse, architect, together with the names of the construction teams and committees. The monument was designed to occupy an axial position in the tunnel approach but is no longer in its original position.

Queensway Tunnel Liverpool
Entrance to Mersey Tunnel, Old Haymarket

Entrance to tunnel, retaining walls and Lodges. 1925-34. Sir Basil Mott and J.A. Brodie with Herbert J. Rowse as architect. Portland stone. Originally an axial and symmetrical design now obscured by subsequent alterations to the layout. Lodges in the form of triumphal arches to left and right of principal axis and retaining walls to the entrance of the tunnel, which is a broad segmental arch. Retaining walls and tunnel entrance have some Art Deco ornamentation and sculpture now partly hidden. The lodges are tall and cubic with fluted buttresses and more Egyptian Art Deco ornamentation.

Mersey Tunnel Entrance, New Quay

Tunnel entrance. 1925-34. Sir B. Mott and J. Brodie Engineers with H. J. Rowse as architect. Portland stone. Tunnel month has decorative plaque over. Long curving wall to right has reeded capping and ends with plaque with winged wheel motif. Iron lamps standards along wall. Wall to left is shorter, with no features of interest.

New Quay Ventilation Station, Fazakerley Street

Tunnel ventilation system. 1925-34, Sir Basil Mott and J.A. Brodie with Herbert J. Rowse as architect. Built as part of the Mersey Road Tunnel. Brick block with two framed entrances and bands of brick decoration at top and bottom, and three ventilation slits. Two towers to rear with decorative brickwork and copper coving. Original doors.

North John Street Ventilation Station

Ventilation station. 1925-34, Sir Basil Mott and J.A. Brodie with Herbert J. Rowse as architect, built as part of Mersey Road Tunnel. Portland Stone. Enormous tower-like structure with design emphasising mass and vertical lines, housing giant ventilating fans. Set on massive rectangular base having windows in vertical bands with Art Deco ornamentation set in horizontal bands above. Tower contains enormously tall slender blind window flanked by attached reeded columns.

George's Dock Ventilation and Central Control Station of the Mersey Road Tunnel, George's Dock Way

Ventilation Station and offices. 1925-34. Sir Basil Mott and J. A. Brodie with Herbert J. Rowse as architect. Built as part of the Mersey Road Tunnel. Portland stone. Storeys. Rusticated ground floor. Windows to north and south in tall recesses flanked by relief sculptures. West facade has windows to end bays, and entrances; centre block has entrance flanked by niches containing bronze figures, term over. East facade has three empty niches, paved areas to north and south have retaining walls, rails and lamp standards, two square piers to north have rusticated bases and banded caps.

Queensway Tunnel Film Credits

The entrance to the Kingsway Tunnel is used as the basis of a tunnel entrance in the video game Grand Theft Auto III—during the 1990s, several members of the game's development team had worked for the Merseyside-based development company Psygnosis.

The tunnels have also been featured in a number of major feature films featuring the Queensway Tunnel (Birkenhead):

Violent Playground (1958), Used Mersey Tunnel as location. Struggle between a Liverpool Juvenile Liaison officer and a young and dangerous pyromaniac.

Harry Potter and the Deathly Hallows (2011) Part 1 (enters the Dartford Tunnel and interior action sequences on motorbike is Queensway). This provided Claire House children's hospice in Wirral with a £20,000 windfall, the money being paid to Merseytravel by Warner Brothers, the makers of the film, for use of the Tunnel as a location for: Fast & Furious 6 (2013) as part of a vehicle chase scene in London, in which there are a number of chase scenes and vehicles splitting off to an alternative exit (Liverpool docks exit) along with a spectacular crash when a car is catapulted through the air and lands on its roof, and Jack Ryan's Shadow Recruit (2014) as the Manhattan Tunnel to which Jack Ryan tracks the real terrorist plot, is the entrance to the disused Queensway Tunnel linking Birkenhead to Liverpool.

It has been said by filmmakers that this is a great location and with enough notice the Tunnel Authority can close the tunnel and divert traffic to the Kingsway Tunnel and therefore not severely inconvenience commuters. As the Birkenhead Tunnel is four lanes so filming an action sequences on two lanes (with traffic going in opposite directions) and put the unit (camera and crew) safely in the other two lanes, is always a good option in today's breathtaking chase sequences.

Chapter 1
History of Tunnel Construction

What Goes into Producing the Tunnel and its Design?
A tunnel is a horizontal passageway located underground, which has been created by the process of excavation to allow traffic or water to travel from two set points. There are many different ways to excavate a tunnel, including manual labour, explosives, rapid heating and cooling, tunnel boring machines or a combination of these methods.

Control of the environment is essential to provide safe working conditions and to ensure the safety of passengers after the tunnel is operational. One of the most important concerns is ventilation – a problem magnified by waste gases produced by trains and vehicles using the tunnel. Clifford Holland addressed the problem of ventilation when he designed the tunnel that bears his name. The Clifford Milburn Holland Tunnel (more commonly known as the Holland Tunnel). This is a road tunnel under the Hudson River connecting Interstate 78 on the island of Manhattan in New York City, and Route 139 in Jersey City, New Jersey on the mainland of the United States. The tunnel was originally known as the Hudson River Vehicular Tunnel or the Canal Street Tunnel; it was the first of two vehicle tunnels, the other being the Lincoln Tunnel.

His solution to the problem of ventilation and removal of the excess vehicle omissions was to add two additional layers above and below the main traffic tunnel. The upper layer clears exhaust fumes, while the lower layer pumps in fresh air. Four large ventilation towers, two on each side of the Hudson River, house the fans that move the air in and out. Eighty-four fans, each 80 feet in diameter, can change the air completely every 90 seconds. The Holland tunnel is the reason the Queensway has its ventilation system today.

The opening of the tunnel is a portal, and the roof of the tunnel, or the top half of the tunnel, is the crown and the bottom half is the invert. The basic geometry of the tunnel is a continuous arch. Because tunnels must withstand tremendous pressure from all sides, the arch is an ideal shape. Tunnel engineers, as with bridge and motorway engineers, must also consider the forces interacting to produce equilibrium on structures.

- Tension, which expands, or pulls on, material
- Compression, which shortens, or squeezes material
- Shearing causes parts of a material to slide past one another in opposite directions
- Torsion, which twists a material

The tunnel must oppose these forces and in order to remain static, tunnels must be able to withstand the loads placed on them. Dead load refers to the weight of the structure itself, while live load refers to the weight of the vehicles and people that move through the tunnel. All this has to be considered after the large amounts of geological and other surveys undertaken before we reach the planning stage.

Almost every tunnel is a solution to a specific challenge or problem and can be described as an underground passage from point A to point B. The tunnel may be used for foot or vehicular traffic, railways or in some instances a canal or a canal aqueduct to supply water for consumption or for hydroelectric stations or sewers. Even cities, with little open space available for new construction, can be an obstacle that engineers must tunnel beneath to avoid.

Often, a single tunnel will pass through more than one types of material or encounter multiple hazards. Good planning allows engineers to plan for these variations right from the beginning, decreasing the likelihood of an unexpected delay in the middle of the project.

Once engineers have analysed the material that the tunnel will pass through and have developed an overall excavation plan, construction can begin. The tunnel engineers' term for building a tunnel is driving, and advancing the passageway which can be a long, tedious process that requires blasting, boring and digging by hand.

The current Crossrail extension of the London Underground scheme is to extend the service of the London Underground and greatly increase capacity on the central London section to accommodate more frequent and longer trains.

A tunnel project starts with a comprehensive investigation of ground conditions by collecting samples from boreholes and by other geophysical techniques. An informed choice can then be made on which way would be best to proceed and which machinery and methods are to be used. In planning the route, the horizontal and vertical alignments will make use of the best ground and water conditions.

Workers generally use two basic techniques to advance a tunnel. In the full-face method, they excavate the entire diameter of the tunnel at the same time. This is most suitable for tunnels passing through strong ground or for building smaller tunnels. The second technique is the top-heading-and-bench method. In this technique, workers dig a smaller tunnel known as a heading or pilot hole. Once the top heading has advanced some distance into the rock, workers begin excavating immediately below the floor of the top heading, this is a bench method. One advantage of the top-heading-and-bench method is that engineers can use the heading tunnel to gauge the stability of the rock before moving forward with the project. Tunnels through mountains or underwater are usually worked from the two opposite ends of the passage. In long tunnels, vertical shafts may be dug at intervals to excavate from more than two points.

Tunnels are constructed in all manner of materials and ground situations from the simple tunnel through a hill to more complex ones through various types of materials varying from soft clay to hard rock. The method of tunnel construction depends on such factors as the ground conditions, the ground water conditions, the length and diameter of the tunnel drive, the depth of the tunnel, the logistics of supporting the tunnel excavation, the final use and shape of the tunnel and appropriate risk management.

To prevent cave-ins from happening, engineers used a special piece of equipment called a shield. A shield is an iron or steel cylinder literally pushed into the soft soil. It supports the surrounding earth while workers remove debris and install a permanent lining made of cast iron or precast concrete. When the workers complete a section, jacks push the shield forward and the process is repeated.

Marc Isambard Brunel, invented the first tunnel shield in 1825 to excavate the Thames Tunnel in London. This shield comprised of 12 connected frames, protected on the top and sides by heavy plates called staves. Each frame was divided into three workspaces, where diggers could work safely. A wall of short timbers separated each cell from the face of the tunnel. A digger would remove a timber section and carve out three or four inches of clay and replace the timber section. When all of the diggers in all of the cells had completed this process on one section, powerful screw jacks push the shield forward.

In 1874, Peter M. Barlow and James Henry Greathead improved on Brunel's design by constructing a circular shield lined with cast-iron segments. The design was first used to excavate a second tunnel under the Thames for pedestrian traffic. In 1874, the shield was used to excavate the London Underground, the world's first subway. Greathead further refined the shield design by adding compressed air pressure inside the tunnel. When air pressure inside the tunnel exceeded water pressure outside, the water stayed out. Soon, engineers in New York, Boston, Budapest and Paris had adopted the Greathead shield to build their own subways

Tunnelling through hard rock almost always involves blasting. Workers place explosives via drilled holes in the rock, which depending on the type of rock will normally be around 10 feet deep. Workers pack explosives into the holes, evacuate the tunnel and detonate the charges. After vacuuming out the noxious fumes created during the explosion, workers can enter and begin carrying out the debris, known as muck, using carts. Fire-setting is an alternative to blasting. In this technique, the tunnel wall is heated with fire, and then cooled with water. The rapid expansion and contraction caused by the sudden temperature change causes large chunks of rock to break off.

The stand-up time for solid, very hard rock may measure in centuries. In this environment, extra support for the tunnel roof and walls may not be required. However, most tunnels pass through rock that contains breaks or pockets of fractured rock, so engineers must add additional support in the form of bolts, sprayed concrete or rings of steel beams. In most cases, they add a permanent concrete lining.

Tunnelling through soft rock and tunnelling underground require different approaches. Blasting in soft, firm rock such as shale or limestone is difficult to control. Instead, engineers use Tunnel Boring Machines (TBMs), to create the tunnel. TBMs are enormous, multimillion-dollar pieces of equipment with a circular cutting plate on front. The circular plate is covered with disc cutters – chisel-shaped cutting teeth, steel discs or a combination of the two. As the circular plate slowly rotates, the disc cutters slice into the rock, which falls through spaces in the cutting head onto a conveyor system. The conveyor system carries the muck to the rear of the machine. Hydraulic cylinders attached to the spine of the TBM propel it forward a few feet at a time.

TBMs don't just bore the tunnels – they also provide support. As the machine excavates, two drills just behind the cutters bore into the rock. Then workers pump grout into the holes and attach bolts to hold everything in place until the permanent lining can be installed. The TBM accomplishes this with a massive erector arm that raises segments of the tunnel lining into place.

There are some basic types of tunnel construction other than a TBM in common use and they have been briefly explained below. This is to give the reader an idea as to the unforeseen ways that would be selected and how the tunnel is constructed.

Cut-And-Cover

This is a simple method of construction for shallow tunnels so a trench is excavated and roofed over with an overhead support system strong enough to carry the load of what is to be built above the tunnel. This may be a simple urban roadway or motorway to a large multi-story building.

This method of construction can be broken down further into

Bottom-Up Method:

A trench is excavated, along with any necessary ground support and the tunnel is constructed in the trench. The methods of construction may be one of many options including in-situ concrete, precast concrete, precast arches, or corrugated steel arches. The trench is then carefully back-filled and the surface finish is then completed to either hide or mask the finished article.

Top-Down Method:

Side support walls and capping beams are constructed from ground level contiguous bored piling or other similar methods. A shallow excavation allows the contractor to install the roof of the tunnel which would be concrete or steel. This method would allow the continuous flow of traffic in high volume areas and minimise any disruption. The main tunnel is then excavated whilst other works are carried out in the main construction area, thus allowing the project to complete quickly and without any undue inconvenience.

Clay kicking

Regularly used in Victorian civil engineering, this method was used for the renewal of the UK's ancient sewerage systems. Clay-kicking is a specialised method of tunnel construction that was developed in the United Kingdom and sees the manual digging of the tunnel in strong clay soil. Unlike previous manual methods of using mattocks which relied on the soil structure to be hard, clay-kicking was relatively silent and hence did not harm soft clay-based structures. The method would see the tunnellers lying on a plank at a 45-degree angle away from the working face. He then inserts a tool with a cup type of rounded end with his feet. Turning the tool with his hands, he extracts a section of soil, which is then placed on the waste removal system, usually a conveyor.

During the First World War, the system was successfully deployed by the Royal Engineer tunnelling companies to quietly deploy large amounts of explosive beneath enemy German Empire lines.

Shafts

A shaft is sometimes necessary for a tunnel project and normally has concrete walls to become a permanent structure and can be used for escape stairways or future maintenance access.

In large diameter shafts, a TBM can be lowered to the bottom and start the tunnel excavation process. Sometimes if a tunnel is going to be long, multiple shafts at various locations will be bored so that entrance into the tunnel is closer to the unexcavated area.

Other Key Factors

Knowing the amount of time, a tunnel will support itself without any added structures is known as 'Stand Up time' which allows the engineers to determine how much can be excavated before support is needed. The longer the stand-up time is the faster the excavating will go. Generally certain configurations of rock and clay will have the greatest stand-up time, and sand and fine soils will have a much lower stand-up time.

If Ground Water is leaking into the tunnel stand-up time will be greatly decreased so it is very important that this is controlled. If there is water leaking into the shaft it will become unstable and will not be safe to work in. One of the most effective ways of stopping this is by ground freezing. Pipes are inserted into the ground surrounding the shaft and cooled until they freeze the ground around each pipe until the whole shaft is surrounded frozen soil. Any water is then pumped out of the tunnel.

Tunnel shape is very important in determining stand-up time. The force from gravity is straight down on a tunnel, so if the tunnel is wider than it is high it will have a harder time supporting itself, and thus decreasing its stand-up time. If a tunnel is higher than it is wide, the stand-up time will increase making the project easier. Everyone knows the story of the egg sitting under a vehicle without breaking. The point load on the egg is spread down the entire egg and out, thus withstanding the load.

The Future of Tunnelling will be an ever-increasing area given new methods of construction and materials to use. Engineers will continue to build longer and bigger tunnels. Recently, advanced imaging technology has been available to scan the inside of the earth by calculating how sound waves travel through the ground. This provides an accurate snapshot of a tunnel's potential environment, showing rock and soil types, as well as geologic anomalies such as faults and fissures.

While such technology promises to improve tunnel planning, other advances will expedite excavation and ground support. The next generation of tunnel-boring machines will be able to cut 1,600 tons per hour. Engineers are also experimenting with other rock-cutting methods that take advantage of high-pressure water jets, lasers or ultrasonic systems. Chemical engineers are working on new types of concrete that harden faster because they use resins and other polymers instead of cement.

With new technologies and techniques, tunnels that seemed impossible even 10 years ago, suddenly seem a possibility. One such tunnel is a proposed Transatlantic Tunnel connecting New York with London. The 3,100-mile-long tunnel would house a magnetically-levitated train travelling 5,000 miles per hour. The estimated trip time is 54 minutes – around seven hours shorter than an average transatlantic flight. Another future option is one from Gibraltar (Southern Spain) to North Africa.

QUEENSWAY TUNNEL

Chapter 2
Queensway Tunnel – The Beginning

For the first five centuries, Liverpool struggled to overcome its geographical problems set her by Mother Nature. Liverpool or muddy pool as it was then known after a charter of the 1190s and referring to a small muddy inlet (hence the name muddy pool). Up till the 1500s, it was no more than agricultural land and villages and not the large urban area we know today.

Through the ages, it fought for a living until on August 28 1207, King John signed a charter founding the town of Liverpool. This charter is proudly kept by the city today in acknowledgement of its first beginnings. The charter, written in Latin translates to

John, by the Grace of God King of England, to all his faithful people who have desired to have burgages (Land Tenancies) *at the township of Liverpool that they shall have all liberties and free custom in the town of Liverpool which any borough by the sea has in our land...*

King John's main attraction for the town of Liverpool was its close proximity for his departure to Ireland and as such, he built a castle which was completed around 1237.

The idea of a tunnel and travelling under the Mersey, not to mention other rivers throughout the world, is a bold draft of an idea from those with a large imagination. The translation of an idea to the realisation of this concrete tube that people use to traverse through is more complicated than just simply digging a long horizontal hole. The tunnel or underground passage would completely enclose the vehicle from entry to exit.

This transition starts with, the ideas being turned into drawing of the base plans, geometric date sourced and collated, final plans drawn, the actual construction process, and finally the finished tunnel. Although not an exhaustive list, these are just the basic items for the process of building a tunnel. Tunnels can even allow a good easy route for cables and pipelines, especially with today's electronic and mobile phone requirements of today's business and pleasure requirements.

The conception of the scheme was a bold piece of a person's imagination. So bold was the idea of a tunnel under the River Mersey, it remained an abstraction for well over a century. It remained something to dream and talked about, (much like the Channel Tunnel and the proposed tunnel under the Atlantic to America, or a crossing over the Straits of Gibraltar) but not for many long years for financiers, administrators and men of science to approach as a practicable enterprise. The birth of the tunnel had to wait until the local transport network was full to bursting point along with the accumulations of wealth and population, and its men of vision and resource.

Rivers are seen as barriers as well as motorways of the sea. As the deep-water haven the River Mersey has made it possible for the great port of Liverpool to grow beyond its small beginnings. Even further down the river at Runcorn, and the Manchester Ship Canal, the sea remained as a great gulf between Lancashire and Cheshire and between Liverpool and the Wirral Peninsula. The only method of travel before the ferries was a simple little rowing boat.

On 20th January 1886, the Mersey Railway opened and started its full public service on 1st February of the same year. This new transport link saw the coming of close links between Liverpool and Birkenhead. The next step was to get a good transport link to allow trading to progress to the next level and link the sea borne goods from both sides of the river.

As early as 1869, Parliamentary powers were obtained for the construction of a tunnel, and those powers ultimately proved to be the original seed to be sown for the future of river crossings though many years passed before the completion of the works. In the meantime, two other Acts empowering the construction of road tunnels were passed, but it was the railway project that found favour with the investing public, and the competitive schemes for horse-drawn vehicles were abandoned.

In 1892, Mr Gladstone, prophesied that in the not too distant future the River Mersey would be pierced a second time. There was no movement in that direction for a period of thirty-three years that his prophesy came to fruition. That historic occasion in 1925, was when Princess Mary started the pneumatic drills on the Pierhead that moved the first spade full of what would be million tons of rock underneath the River Mersey.

In 1914, a committee was put together to study the problem of cross-river traffic, was already becoming acute owing to the increased number of motor vehicles. The outbreak of World War One demanded attention to sterner business, and yet about half-way through the war, in 1916, the late Sir William Forwood, in a newspaper discussion strongly advocating a tunnel as being less obstructive and less costly than a bridge.

The end of the war brought such an enormous development due to the plethora of surplus war vehicles available. The ferries were no longer able to cope with the streams of vehicles, carts and people wanting to cross the river. This popularity of the ferries resulted in long queues on both sides of the river. It showed that the existing facilities were completely inadequate, and influential protests were made on the ground of lost trade and profit for the commercial trade in the area. Mumblings were getting louder and louder in wanting a resolution to the problem as soon as possible.

Finally, in 1921, Sir William Forwood, called the Merseyside community together and his pleading to the Liverpool Chamber of Commerce brought the problem to a proposed solution. The plan that was favoured was for the construction of twin tunnels between Liverpool and Birkenhead. One tunnel would be for fast and the other for slow traffic. This new tunnel would cost about £2,000,000; (around £77,950.20) and it was suggested that the work should be undertaken by a public company to be guaranteed by the four boroughs of Liverpool, Birkenhead, Seacombe, and Bootle.

Now that the war was over and there was an abundance of the new motor vehicles around, it quickly became apparent that the need for a bridge over or tunnel under the River Mersey would be needed. Up until this time, there was only the ferry system to transport people, carts and goods across the river and this was fast becoming a very compact process with the additional traffic week in week out. This system also had the problems of carts and other vehicles backing up close to the terminals waiting to board the next available ferry.

In January of 1922, Sir Archibald Salvidge tabled a motion for the appointment of a committee to report on the feasibility of a scheme of as to the improvement of transport facilities across the river. This would be by the provision of a bridge or tunnel. It included a provision to engage the services of an eminent engineer to investigate and report on the proposals. The possibility of an adequate improvement of the ferries having also been practically ruled out, so a unanimous adoption of the motion was the first practical step towards the realisation of a modern and a crossing for the river which would see the joining of the national motorway beneath the waters of the Mersey.

An invitation to Birkenhead, Wallasey (both now part of Wirral Council) and Bootle (now part of Sefton) Councils to appoint the Merseyside Municipal Coordination Committee. This would consist of six area representatives and be chaired by Sir Archibald Salvidge. Throughout 1922, the committee deliberated on the proposals and, towards the end of June, had decided on the two engineers. These were to be Sir Maurice Fitzmaurice and Sir Basil Mott, (whose company still trades today as Mott MacDonald) to study the problem in consultation. They were to liaise with Mr John A. Brodie, the then Liverpool City Engineer on the merits of the two options.

Finally, after some 12 months, a large report was read and presented to the Co-ordination Committee at the end of September 1923. The report was detailed and included elaborate illustrations and maps of the proposals. The report was given that the docks on both sides of the river were exceptionally busy and the bridge would have to be of such a height to allow ships to pass under it. This single fact, amongst others, would prove to be its downfall. The bridge would have spanned from the old castle site to Woodside, with a central span of 2,200 ft, a clear headway of 185 ft, and a width of 90 ft.

The cost of the bridge, without a branch off to Wallasey would be in the region of £10.5 million (£389,751,000.00). The unanimous decision was for a tunnel under the river which would be in the region of £6.5 million (£253,338,150.00) and would be a double deck construction with two branches on either side of the river. These would-be Birkenhead and Seacombe on the Wirral side, and Liverpool City and Bootle on the Liverpool side, both of which would allow full access to the docks.

The recommendations were for a tunnel under the river of 44 ft (13.41 metres) internal diameter, starting from a shaft to be sunk on a site near the dock board offices on the Liverpool side. This would pass under the river clear of the Mersey Railway tunnel, to the Birkenhead side, south of the Seacombe Landing Stage.

On the Liverpool side two branches were proposed as exits – one for heavy dock traffic, with an open approach to New Quay, and one for light, fast traffic and tramway services, with an open approach to Whitechapel in a proposed new, wide street. On the Wirral side, one branch was to serve Birkenhead with an open approach near the Woodside Hotel, opposite the Woodside Ferry Terminal, (demolished due to fire in 2011/12) and another to serve Wallasey.

In this, the original scheme, and the reason that the great diameter of 44 ft (13.41 metres) was provided for four lanes of traffic on the upper deck and for two sets of tram rails on the lower. Provision was included for ventilation by means of inlets for fresh air under pressure, and for outlets for exhausting the vitiated atmosphere.

The main figures of the estimated cost as at 1921:

Tunnel works	£5,410,000 (£210,855,291.00)
Property and easements	£570,000 (£22,215,807.00)
Parliamentary, legal and Engineers' expenses	£420,000 (£16,369,542.00)
Grand total	**£6,400,000 (£249,440,640.00)**

The main Co-ordination Committee submitted the report to a special sub-committee, consisting of two representatives from each of the municipal councils. Under the terms of reference, the sub-committee was 'to consider the report, and particularly its financial aspect, in close detail,' The sub-committee were to look at the possibility of improving the cross-river facilities, and subsequently to submit a definitive proposal with financial details.

The committees knew there was a lot riding on their proposals and decisions. There were a large number of men unemployed, the iron and steel industry was on its knees. The proposals had gone countrywide and rumour has spread that there was to be a great new north – south connection. This was the age of the birth of the motor car and any additional roads and drivable surfaces were gladly accepted.

The autumn of 1923 and the whole of 1924, discussions took place in both private and public on the pros and cons of the great project. The discussions looked at the opinions expressed by the several boroughs, and of the nation itself. Finally, Mr Sidney Dawson criticised the heavy cost of the scheme. The tunnel was agreed and a protest was made by the Mersey Docks and Harbour Board to the Seacombe branch of the tunnel, on the ground that it might impede dock developments on that side of the river.

On the other hand, Sir Archibald Salvidge championed the scheme with characteristic resolution and thoroughness and he had the valuable backing of Sir William Forwood and of the local members of Parliament, including the ministry of transport. Sir Archibald Salvidge, who with Sir George Etherton (Town Clerk of Liverpool) and of Mr Arthur Collins, the financial expert, whose estimates formed the basis of the claim towards a substantial Government contribution.

Not long after the tunnels were complete and in use, people of both sides of the river realised the importance of the tunnels. This was particularly so for the people of the industrial Liverpool. The rural settings of Wirral became an advantageous area to visit for beaks. Tour companies organised a 'Charabang' (Coach) trips to Wirral, particularly during the holiday periods. Now most of the traffic through the Tunnels is from people going to work in Liverpool from Wirral and the surrounding areas, with the corresponding traffic coming from Liverpool at the end of the working day.

Chapter 3
The Tunnel Makers

Tunnel Makers

Sir Basil Mott

Sir Basil Mott, Bt., C.B.F.R.S, (1859–1938) appointed as the engineer-in-chief by the Tunnel Committee. Sir Basil was educated at both the International College in Switzerland and Royal School of Mines. During his time at the Royal School of Mines, Sir Basil became Murchison medallist in 1879. Sir Basil was also chairman of the works committee responsible for the restoration of St. Paul's Cathedral.

Sir Basil served on many Government committees during the First World War. He was made a Companion of the Order of the Bath in 1918, and created a baronet in 1930. He was a past-president of the Institution of Civil Engineers, an Associate of the Royal School of Mines, and a Fellow of the Royal Society and of the Imperial College of Science and Technology.

Mr John A. Brodie

John A. Brodie, was a president of the Institution of Civil Engineers, and associated with Sir Basil Mott for several years. He was employed by the Liverpool Corporation for thirty-six years, most of which he held the position of city engineer.

As a planner and motorways engineer, he was prominent in some of the major road schemes throughout Liverpool. This work included Queen's Drive, the bridges over the old George's Dock, Renshaw Street widening, Scotland Road. The Otterspool scheme which included some of the spoil from the Mersey Tunnel.

Mr B. H. M. Hewett

Bertram Henry Majendie Hewett was, from 1926 until his death in 1933, engineer-in-charge under Sir Basil Mott. He had responsibility for the work of tunnelling, but also for the ventilating machinery, the electric lighting system, and the auxiliary plant. All of this entailed an immense amount of detailed study and attention to detail.

Mr David Anderson

Sir David Anderson (1880–1953) was a Scottish civil engineer and solicitor. Sir David Anderson was born in 1880 at Leven, Fife, Scotland. In 1921, on his return from Army service, were he attained the rank of Captain in the Royal Engineers. Mr Anderson joined a partnership with fellow engineers Basil Mott and David Hay, forming the company Mott Hay and Anderson. Mott, Hay and Anderson traded until 1989, when it merged with Sir Mott. MacDonald & Partners to form Mott MacDonald. Mr Anderson was elected president of the Institution of Civil Engineers for the November 1943 to November 1944.

Mr C. B. H. Colquhoun

Mr C. B. H. Colquhoun, B.Sc., A.M. Inst. C.E., the resident engineer, succeeded Mr Hewett as the representative of Sir Basil Mott on the tunnel works. Educated at King's College, London, he was first employed by Messrs. Muirhead, Macdonald, Wilson and Co.

Following a short stay of work across the Atlantic, and upon his return to England, in 1930, Mr Colquhoun was appointed resident engineer of the Mersey Tunnel under Mr Hewett. In November, 1933, he became resident engineer-in-charge, and in that capacity, has superintended the finishing touches necessary to the completion of the great enterprise.

Mr Herbert J. Rowse

Mr Herbert J. Rowse, F.R.I.B.A. (1887–1963) the architect to the Mersey Tunnel Joint Committee since 1931, was a native of Liverpool, and was educated at the University of Liverpool. Winning the Holt travelling scholarship he studied architecture in France and Italy, and travelled extensively in Canada and the United States of America. He holds what is probably the unique record of having won the first premium in all the competitions for which he has entered since the war.

These included the India Buildings, Liverpool (in partnership), new clubhouse at Heswall Golf Club, and the new headquarters for Martins Bank, Liverpool. Martins Bank and India Buildings have made striking and architecturally revolutionary changes in the varied profiles of the city's skylines.

It was doubtless because of his work in Liverpool that the tunnel committee appointed Mr Rowse as their architect for the ventilation buildings. Here the problem was to coordinate the complicated engineering requirements so as to provide the most efficient disposition of the plant, and at the same time produce economical buildings of aesthetic merit. The successful outcome is now apparent in the six ventilation buildings Mr Rowse was also called upon to design and advice upon the architectural treatment and the equipment of the interior of the tunnel itself. Of the four great entrances in Liverpool and Birkenhead, including the lighting shafts, tollbooths, lamps, pylons, his designs stand true today as they did in their conception.

Professor John Scott Haldane

Dr John Scott Haldane (1860–1936) was thought of as one of the greatest living authorities on ventilation and on air analysis. In the course of his researches he exposed himself several times to serious dangers. He descended an old mine with Sir William Atkinson, who became Chief Inspector of Mines for South Wales, to observe the effect of carbon dioxide poisoning on the human body. That experiment led to important discoveries in the functions of the respiratory system to study the effects of carbon monoxide poisoning, which was responsible for the majority of deaths in colliery explosions.

Professor Haldane was made a Companion of Honour in the King's Birthday Honours List of 1928 for scientific work in connection with industrial disease. Born in Edinburgh in 1860, Professor Haldane was educated in medicine and in those branches of science in which he has become so outstanding a figure there, and in the University of Jena. He is an honorary professor and the director of the Mining Research Laboratory, Birmingham University. In addition to scientific work which has won him honours from many parts of the world besides England, he invented a remarkably simply-operated apparatus for tracing the presence of minute percentages of methane gas in air, which, like his self-contained rescue apparatus, has been of a value not to be calculated in mining work.

Professor Douglas Hay

Professor Douglas Hay, (1889–1949) started practical work in mining when he was apprenticed at the Shirebrook Colliery, Notts, and he gained the first-class colliery manager's certificate in 1910. Official appointments in coalmines in England testify to his distinction in his profession. He was H.M. Inspector of Mines in Durham in 1913 and 1914, and, after active service in France as a major in the Royal Field Artillery, and with the Royal Engineers, Field Survey, during which time won the Military Cross, he was again appointed Inspector of Mines in Durham and in North Staffordshire, from 1919 to 1922.

Mr P.J. Robinson

The tunnel engineer's only adviser on electrical matters has been Mr Percival James Robinson, Ailing, M. Inst. E.E., M.I. Mech. E., Liverpool City Electrical Engineer, whose rise to eminence in his profession has all been achieved in the service of the city, and is acknowledged throughout the country.

Mr Robinson was born in Kent, in 1879, and was apprenticed to the electrical profession. He studied at Finsbury Technical College. The most striking work that Mr Robinson carried out was the design of the Clarence Dock Power Station, Liverpool, which at the time was officially stated to be the most efficient in the country. He was a member of both the National and the North-West England Consultative Committees of the Central Electricity Board. Liverpool University, in 1933, gave him the honorary degree of Master of Engineering, and he is an honorary life member of the Society of Civil Engineers of France.

Mr E.W. Monkhouse

Mr E. W. Monkhouse, M.A., M. Inst. C.E., M.I. Mech. E., M.I.E.E., acted as adviser to Sir Basil Mott in connection with the problems arising out of the insulation of the machinery and plant in the ventilation stations, to guard against the transmission of noise and vibration. Mr Monkhouse was a partner in the firm of Messrs. Burstall & Monkhouse, of London.

Messrs. E. Nuttall, Sons & Co Ltd

Messrs. Edmund Nuttall, Sons & Company, Limited was awarded seventy per cent of the construction of the work for the Queensway Tunnel. Prior to 1925, the company had had a long experience of civil engineering work, and particularly of tunnelling and dock construction.

The company, one of the earliest to build structures of reinforced concrete in England, were the contractors for the Royal Liver Building, Liverpool, designed by Walter Aubrey Thomas which was by far the largest and heaviest structure attempted in that material up to that time.

This contract lasted from 1908–1911 and £800,000 (£31,180,080.00 at its opening in 1911) in value proved a turning point in the history of the company, whereby it grew up from what was then a middle-sized firm operating mainly in the North of England to its present stature as one of the major civil engineering contractors in the United Kingdom and of international repute.

The New Mersey Tunnel (Wallasey Kingsway) were started in 1967 when the Birkenhead (Queensway) tunnel became inadequate for the traffic it generated, once again this important work was entrusted to Edmund Nuttall. The year of 1972 saw a joint venture, equal partnership between Edmund Nuttall, Guy F. Atkinson Co, Balfour Beatty & Co Ltd and Taylor Woodrow Construction Ltd to form Cross Channel Contractors. In 1973, Cross Channel Contractors were awarded Phase II Works on the Channel Tunnel.

In 1978, the company was bought by Hollandsche Beton Groep (later HBG). In 1980, Edmund Nuttall built the first post-tensioned concrete road bridge to be constructed over the River Tyne, The New Redheugh Bridge. In 1979, Nuttall acquires Mears followed by Hynes Construction in 1992, John Martin Construction in 1999, and Finchpalm Ltd in 2000. In 2002, HBG was acquired by Royal BAM Group.

Sir Robert Mcalpine and Sons

The firm of Sir Robert McAlpine & Sons was founded by Sir Robert McAlpine. Sir Robert McAlpine is still a leading UK building and civil engineering company. Recent high-profile projects include the O2 Arena, Emirates Stadium, the Eden Project, Cabot Circus in Bristol, the M74 Completion in Glasgow and the Olympic Stadium.

Professor P. G. H. Boswell

Among the various professionals, in the early days of the enterprise, the promoters, engineers, and the Committees of Parliament relied was Prof Percy G. H. Boswell, D.Sc. (bond), the eminent geologist, at that time George Hardman Professor in Geology in the University of Liverpool.

Born in Woodbridge, Suffolk, Prof Boswell, who since leaving Liverpool in 1930 has occupied the chair of geology in the University of London (Imperial College of Science and Technology). He had a distinguished career as a teacher, administrator and writer on geological and other scientific subjects. During the war, he was scientific adviser to the Ministry of Munitions, and was awarded the O.B.E.

Mr David Hay, M. Inst. C.E

Mr David Hay, who is a member of the firm of Messrs Mott, Hay & Anderson, has been closely associated with 'Sir Basil Mott in the responsibilities imposed upon him by this great engineering task.

The Sturtevant Engineering Co Ltd

The Sturtevant Engineering Co, Ltd, were the pioneers in this country of mechanical ventilation, and owing to the very wide experience gained made them among the leading people in this branch of engineering. Although, a purely British concern, they were for many years closely associated with the B.F. Sturtevant Company, of America, who installed ventilating plants in the Holland Tunnel under the Hudson River, at New York. This tunnel will come up again in the history of the ventilation shafts. Much of the technical data regarding these installations was available and applicable to the special conditions obtaining in the Mersey Tunnel.

Messrs. Walker Bros. (Wigan), Ltd

Messrs. Walker Bros. (Wigan), Ltd, who have supplied 10 fans with a total capacity of 2,850,000 cubic feet, are an important firm of engineers who have distinguished themselves by their work in connection with the ventilation of mines and tunnels. The firm was founded in 1868 by Mr J. S. Walker. The company then came under the direction of Major J. S. A. Walker, Mr A. C. Walker, Mr R. Barton Walker and Mr Densmore Walker.

Walker Bros. were contractors to the Admiralty and War Office, and they also undertook important government contracts in India and the colonies. They secured the ventilation contract for the Severn Tunnel, which required a fan capacity of 800,000 cubic feet of air per minute, and they have played an important part in the solution of the problems, which arose in connection with the ventilation of the Mersey Tunnel.

Metropolitan Vickers Electrical Company Ltd

The Metropolitan Vickers Electrical Company, of which the chairman is Sir Felix J. C. Pole, is one of the leading electrical manufacturing concerns of the world. The company's works at Trafford Park, Manchester, occupied approximately 160 acres. The apparatus made by the company was well known to Liverpool engineers, as much of the company's plant has been supplied to the Liverpool Corporation. This included two 25,000-kW turbo generators for the Lister Drive Power Station, and two 51,250-kW turbo altimetre sets at the Clarence Dock Power Station.

Mersey Cables, Ltd

Mersey Cables Ltd was entrusted with the contract for the whole of the electric cables installed in the tunnel. The managing director of the company is Mr W. S. Taylor, spent a lifetime in the study of the problems which confront those who are engaged in this branch of electrical engineering. Although, the company was only formed in 1926, the quality of its productions enabled it to secure contracts with government departments and a number of the large municipal authorities.

Chapter 4
Agreement of the Various Parties

After the concept of a tunnel was agreed, all the boroughs involved held numerous lengthy negotiations as to costs and locations of such things as the entrance and exits. As expected, everyone was fighting their own corner and looking after their own residents and commercial interests. As in all such cases, even today, there was bound to be a little bit of give and take on all sides but who was going to blink first and give the most, and who was going to be the eventual victor?

The government wanted the crossing to be free of any tolls, but after several years of negotiations it was agreed that the government would pay half the construction cost, one quarter would come from the rates in Liverpool and Birkenhead and one quarter from tolls for a period of up to 20 years. This was authorised in a 1925 Act and a Mersey Tunnel Joint Committee was formed comprising of Birkenhead and Liverpool Corporations.

In March 1925, both Birkenhead and Wallasey Councils (then part of Cheshire and now part of Merseyside since the Local Government Act of 01 April 1972) Councils held a meeting in their respective town halls and decided to support the tunnel scheme. Birkenhead had one person who opposed the scheme and Wallasey had four people in opposition, but there were conditions placed upon the scheme to protect the interests of people and business within the boroughs.

Birkenhead Council stated they would pay only 4d (£0.88p in £53) in the £1 and had stipulated that the ferries were pooled. This was a major part of the scheme and would make the whole tunnel scheme a much more viable business option, not to mention allow an element of future proofing the whole scheme. Wallasey stated they would pay 6d, (£1.33 in £53) in the £1 but the loss-making goods ferry would be taken over by the Tunnel Administration and the profitable passenger ferries would be retained by the Wallasey Council. This demand was met with severe criticism as it was seen that Wallasey were laying down unfair conditions. However, Wallasey was seen to be in favour of the underground tramway, despite its possible financial loss to the foot fall on the passenger ferry. They stated that they would not wish for this to be seen as an ultimatum, more a case of the borough looking after its residents and business interests.

In the days that followed and to the surprise of everybody, Sir Archibald Salvidge proposed the elimination of both the Wallasey arm and the tramways, and consequently of Wallasey itself. This action was criticised but in the light of events it was probably going to be the inevitable conclusion to the events that were taking place and the demands that were being made.

For one thing, there was no time for further negotiations if a bill was to be ready for the next session of Parliament, but more important were three other considerations, namely

1. Birkenhead's opposition to trams, without which a Seacombe branch would be a waste of money.
2. The saving of £1,500,000 (£79,515,750) in capital cost.
3. The avoidance of a clash with the Mersey Docks and Harbour Board respecting the safety of the Birkenhead dock entrances.

Bootle Council decided to co-operate and to contribute a 2d (£0.44p in £53) in the £1 rate, the decision of the City Council meant that responsibility for the revised scheme was to be shared by Liverpool, Birkenhead and Bootle. On that basis, a Joint Committee to promote a Parliamentary Bill was set up, with Sir Archibald Salvidge as chairman, Alderman R. J. Russell (Birkenhead) as deputy chairman, and Mr George Etherton, who had been appointed Clerk to the Lancashire County Council, as legal adviser.

A financial re-adjustment for the scheme was announced by Sir Archibald shortly after the meetings. The exclusion of the Wallasey branch and the tramways had brought the estimate of cost down to £5,000,000 (£265,052,500.00). The Government now promised £2,500,000, (£132,526,250.00) towards the costs while consenting to an extension of the toll period from 15 to 20 years. The proposed grant for the tramways was withdrawn, but the total value of the 50 per cent money grant and of the capitalised value of the concession in respect of tolls was estimated at £3,467,000 (£183,787,403.50).

This caused some opposition which became sufficiently organised to force a poll of the ratepayers of the three interested boroughs. Sir Archibald Salvidge led a group of pro tunnel campaigners who were influentially supported, irrespective of political party. Civic administrators, merchantmen, industrialists, leading ship-owners, retail traders, and trade-union leaders, all declared that the progress and prosperity of Merseyside must be notably advanced by the construction of the tunnel. Bootle was the only one with mutterings of dissent that could be called ominous. Some of the Labour spokesmen even suggested the view that housing should come first. If the people do not have a job or prospect, how can they afford to live? It is for this reason you have to ask why the tunnel and all its financial prospects would be the better option.

The polls took place in the first week of May. Bootle polled first, and by giving an adverse vote, cut itself adrift from the partnership. Only 22 1/3 per cent of the Bootle electorate troubled to visit the booths. For the Bill were 3,157, and against it, 3,313 majority against, 156. Bootle's verdict undoubtedly proved a stimulus to the electors of Liverpool, who three days later declared for the Bill in the proportion of nearly 12 to 1. The figures were for, 79,906; against 6,937 majority for, 72,969. On the following day Birkenhead by a 4 to 1 majority put the final seal on the scheme. They were for the Bill, 15,359; against, 3,693; majority for, 11,666. Thus, the tunnel became the exclusive enterprise of the two leading Merseyside communities Liverpool and Birkenhead.

Normal practice is to have a bill go through Parliament within a specific time and period. In the case of the Mersey Tunnel, the usual date for lodging private bills was long past, the fact that the Mersey Tunnel Bill was calculated to give a large volume of widespread employment secured for it an entry to Parliament for the ensuing session, through the suspension of the standing orders.

The bill reached the committee of the House of Lords on June 15th, 1925, and after a hearing of thirteen days, emerged successful. Wallasey's claim to compensation for possible loss of goods ferry traffic, and the Birkenhead Union Guardians' plea that the tunnel tolls should be rateable, was disallowed; but the Mersey Railway Company obtained a protective clause prescribing minimum tolls for 'buses and their passengers. These findings were endorsed by a committee of the House of Commons, after an 8-day hearing; and having passed through all the Parliamentary stages, the Bill received the Royal Assent on August 8th.

In view of subsequent developments, in mining and the resulting dangerous gases, Dr John S. Haldane, president of the Institute of Mining Engineers, and an acknowledged authority on gases and ventilation was brought in. Dr Haldane said that he had examined the ventilation proposals, and was satisfied with them. It would be absolutely impossible, he added, for carbon monoxide to gather in pockets in the tunnel, because it would be swept away by fresh air under control at every point along the tunnel. Mr Basil Mott, the designer of the tunnel, and Mr J. B. Lister, of the Sturtevant Engineering Company, also spoke confidently of the adequacy of the ventilation method.

Mr Mott stated that he had allowed £840,000 (£44,528,820.00) for contingencies of the allowance for the ventilation. He was confident that if the tunnel was full from end to end, there would be sufficient ventilation for the tunnel. In accordance with the provisions of the Act, the Liverpool City Council and the Birkenhead Borough Council proceeded at once to constitute a statutory tunnel authority, under the title of the Mersey Tunnel Joint Committee.

Liverpool's representatives, were to be, Aldermen Sir Archibald Salvidge (who later received the Honorary Freedom of the City in recognition of his work in promoting the scheme), Sir Max Muspratt, Sir John Fitting, W. J. Burgess, John Clancy, Thomas Dowd, Austin Harford, and W. A. Robinson (later succeeded by Alderman Luke Hogan), and Councillors Lawrence Holt and Thomas White.

Birkenhead had the following people on the committee, Aldermen A. H. Arkie, W. M, Furnival, R. J. Russell and G. A. Solly, and Councillors W. H. Egan, Luke Lees and Charles McVey. Basil Mott and John A. Brodie were appointed joint engineers, Mr Mott to be the chief engineer and personally responsible to the committee. Mr Walter Moon, the Town Clerk of Liverpool (who had the valuable assistance of Mr W. H. Baines, the Deputy Town Clerk), was appointed clerk to the committee, and Mr F. J. Kirby valuer.

On December 8th, 1925, a contract was given to Edmund Nuttall, Sons & Company Ltd of Manchester, to start construction of the tunnel shafts. The company was instructed via a contract to dig a shaft on each side of the river and to commence the boring of the shafts under the river. The contract was worth £443,079 (£22,957,734.33) and was to last no more than twenty-five months and only local labour was to be used.

16th December 1925 finally saw the start of the tunnelling process was started in a ceremony on the Pier Head when Princess Mary started the drill for the first shaft. She was accompanied by her husband the Viscount Lascetics, Lord and Lady Derby, the Lord Mayor Sir Frederick Bowring, Lady Mayoress Mrs E W Hope and Sir Archibald Salvage along with members of the Tunnel Committee. The ceremony was also in the presence of a large selection of local influential people and the general public.

Sir Archibald Salvidge presented the Princess with a gold key, and with a turn of the wrist set in motion fourteen powerful pneumatic drills placed in position at the bottom of the disused southern portion of the old George's Dock. As she started the drills, the Princess said:

I declare the work started on this great scheme.

This set-in motion a momentous engineering process that was to last until the summer of 1934.

On the 10th of March 1926, a similar but less spectacular ceremony took place near Morpeth Dock, Birkenhead, where Sir Archibald Salvidge, wielding a pick and shovel, had the honour of removing the first sod. As the local population was full of news for this new engineering process that is finally under way, little did they know that there were still a few things to be ironed out. The committee was still procrastinating over the location on both sides of the river for the tunnel entrances. This seems a little odd if they had already started to dig the tunnels, but there was quite a way to go yet in the tunnelling process. The Birkenhead Entrance was to be at Woodside but alternative locations were discussed. The Liverpool entrance saw objections to the suggestion of Stanley Street and Whitechapel. The New Quay site however was a good selection and this was not disputed.

The Committee was at loggerheads as to the decisions on the entrances and it seemed that the resolution was to be as depicted by Parliament. The controversy between Liverpool, Birkenhead and the Tunnel Committee would remain for some time. In Birkenhead, the choice of entrance was to fill the press, public debate and the council for many months. This debate was for a demand for a second entrance near the docks, and the transfer of the main entrance to the neighbourhood of the Haymarket.

On the Liverpool side, the location of the main entrance was also a bone of contention and the Whitechapel entrance would be better if it was moved to the Old Haymarket. This would allow a better focus on traffic and in addition avoid the requirement for the construction of a new road system from Whitechapel to Lime Street at the edge of the city. This new system would add around £1,250,000 (£66,263,125.00) to the cost of the tunnel and where would the money come from?

July saw the Birkenhead Committee reject the majority of the tunnel committee, and passed their own resolution of 40 to 5 votes. This vote demanded an entrance near the Haymarket with a second entrance at the docks and not Woodside. The additional costs of the works were put at £120,000 to £160,000 (£6,361,260.00 – £8,481,680.00). Birkenhead Councils wishes on the entrances were to be respected, notwithstanding the Act of Parliament on the condition that it would be practical.

However, Liverpool Council, Mersey Docks and Harbour Board along with the Ministry of Transport along with the tunnel engineers were placing proposals. They were against the plans and they were rejected by the tunnel committee in July. The tunnel engineers were instructed to proceed as planned and the Birkenhead proposals had no practical solutions.

This now proved to be a tense time for negotiations and would it be the death of the tunnel if no solution could be found? The initial shots were taken by Birkenhead as they issued writs to both the tunnel committee and Liverpool City Corporation. This was to gain a stay of execution for the works on the Woodside site and to compel the promotion of an amended bill to allow Haymarket and the docks to be the chosen entrances.

The compromise saw the exit of three representatives from the tunnel committee who had wanted the Woodside Entrance, so out went Alderman R. 3. Russell, Alderman Arkle and Councillor Luke Lees which saw the following changes Alderman Dennis J. Clarke, who became deputy chairman of the Tunnel Committee, Alderman Purnival, Alderman Solly, Councillors Egan, McLellan, McVey and Van Gruisen.

The extra cost of the new works, along with the new road was put at £374,000 (£19,825,927.00) plus £72,000 (£3,816,756.00) for another link with the Haymarket and Chester Street. Sir Alexander Binnie Son and Deacon were commissioned by the Birkenhead Corporation to cost the improved approach to Chester Street and this came out at £338,000 (£17,917,549.00).

Christmas is a time of good will and it seems this was to be the case for the tunnel. Sir Archibald Salvidge stated that no opposition would be offered to the Birkenhead's Bill. This would be on a condition that Liverpool City Council would not be committing any extra cost. If they were required to commit extra costs, the bill would be fought inch by inch.

The new year of 1927 had arrived with an anxious peace between the parties to the tunnel and Sir Archibald Salvidge offered to arbitrate between the parties. Liverpool declined as they were not convinced that the Woodside option was effectively dead in the water. A vigorous campaign for and against the bill, with supporters of the bill pledged themselves that the 4d (£0.91p) limit would not be exceeded. Alderman Russell, led the opposition, contended that the new proposals, inclusive of consequential street works and re-housing, would cost £736,000, (£39,015,728.00) and involves an addition of Is. 3d (£0.68p) to the rates of the borough.

On January 31st 1927, the vote was cast with a majority of 3,516 voting for a new entrance which was chosen as Rendel Street (Birkenhead Dock Exit). 1,671 voted for the new road which was just under a third of the populous who were eligible to vote. This entrance mainly served the Wirral Docks and those wishing to travel to the resort of New Brighton, which is now reached from the Wallasey Tunnel. The branch has not been used since 1965 when it became apparent that the branch tunnel and main tunnel junction was becoming dangerous to use. There were also unconfirmed stories of policemen and a lollypop man being knocked over.

Over the coming months, discussions took place between all parties with Lord Birkenhead, the then Secretary of State. April 1927 saw the announcement that an agreement had finally been reached. This agreement, Birkenhead has two entrances with the Woodside entrance being set at Bridge Street / Chester Street, with a branch on Rendel Street. The additional costs of this work would be put at £22,000 (£1,166,231.00).

To pay half the cost for the additional works, it was put to the Ministry of Transport that an extension of 25 years be added to the tolls. The remaining half would be paid by the Birkenhead Corporation and the Tunnel Committee, both of whom would pay £55,000 (£2,915,577.50) each. Following on from this agreement and the subsequent additional costs, a peace treaty was signed by the tunnel committee. Lord Birkenhead was asked to be guest of honour a formal Lord Mayor Banquet to honour the agreement.

The revised bill received Royal Assent in July, after its successful passing by Parliament earlier in the year. Across the river in Liverpool, opinion was still divided as to Whitechapel or the Old Haymarket as the main entrance to the tunnel. Whitechapel was quickly becoming the second cousin to the preferred option of the Old Haymarket. Despite numerous voices of concern by the press, city council, the tunnel committee stood its ground. They began work on the excavation from underneath Brunswick Street in the direction of Whitechapel.

In Birkenhead, it was becoming apparent that the Bridge Street proposal was losing favour as it seemed the dock traffic would be separately accommodated. Now that the dock traffic was to be separately accommodated, Woodside had also lost its attractiveness, and the view was held almost unanimously that both the town and the regions beyond would be best served by an entrance in the vicinity of the Haymarket. Ever the peace maker, Sir Archibald Salvidge caused a sensation in November by submitting alternative plans to the Tunnel Committee. After a close study of the previous problems, with the engineers (Messrs. Mott & Brodie), the valuer (Mr F. J. Kirby), and the clerk (Mr 'Walter Moon) it was concluded that Whitechapel was to be abandoned in favour of a new route with a loop at New Quay, constructed under Dale Street to the junction with Manchester Street.

On the Liverpool side this proposal would see a parcel of triangular land owned and earmarked by the council for their new offices, would provide a cutting to permit an exit opposite St Johns Garden and St Georges Hall from the Old Haymarket. On the Birkenhead side, the main tunnel line would avoid the curve backward to Woodside and go beneath Hamilton Square and Albion Street to Market Street South, near the Haymarket, a point giving easy access to the main roads to Chester. However, this proposal would see the demolition of the Carnegie Free Library and the building of a new library in Borough Road. It was stated that this alteration would not add to the costs, due to savings in expenditure on property and cementation works in the tunnel.

The new exists would have a gradient of around 1:30 from the original intended 1:20 and a resulting saving of £1,300,000 (£68,913,650.00) on a new road scheme to connect Whitechapel and Lime Street. The Byrom Street to East Lancashire road was deemed to still be needed and would result in further arguments for the Old Haymarket Entrance. In April 1928, the Mersey Tunnel Act No 3 was passed unanimously by the committee and received its Royal Assent.

During 1928, despite the legal arguments and ill feelings the tunnel was progressing along. It was slowly moving towards a mid-river connection from each side of the river. By the end of March, it was announced that a small crust of river bed rock was the only thing that now separated Liverpool from Wirral. Arrangements were made for the ceremonious break through that has become synonymous today with major tunnel projects. The breakthrough was completed on 3rd April 1928, Sir Archibald Salvidge, the Lord Mayor of Liverpool (Miss Margaret Beavan), the Mayor of Birkenhead (Alderman F. Naylor), Mr Basil Mott and Mr John A. Brodie (joint engineers), Mr Walter Moon (clerk), Mr J. Gibbins, M.P., Mrs. Mercer, an ex-Mayor of Birkenhead, Mr G. O. Lynde (representing the contractors), and members of the Tunnel Committee.

All of the dignitaries were clothed in oilskins, sou'westers, and high gum boots to keep out the constant dripping of water and damp muddy conditions below the river bed. After their carriages, had taken them to the mid river position from their respective ends of the tunnel, they perched themselves onto a makeshift wooden platform to complete the ceremony. Sir Archibald Salvidge was chosen to strike the first blow with a pick and then a sledge hammer, both of which proved unsuccessful.

Later tunnel workers were brought in the use a pneumatic drill to assist in the operation and weaken the rock. Sir Archibald Salvidge tried a second time with the pick and this time was successful with the point of the pick. Load cheers cried out from both sides of the tunnel as it was now realised that Liverpool and Wirral were finally one. All this was complete with an accuracy of around five eighths of an inch (15mm) for both the top and bottom tunnels. A truly remarkable achievement in any age and especially considering they had basic equipment unlike today's tunnellers who use GPS and laser levels to check the buildings above ground and any subsidence. The Crossrail project in London is a good example of this as they monitor the numerous historic buildings above ground for movement 24 hours a day.

Once the initial break through was complete the dignitaries stood back and allowed the tunnel workers to complete the task of finish what they had started. The hole was enlarged to around five feet in diameter to allow the two chief magistrates to shake hands and complete the joining of these two great boroughs 150 feet below the River Mersey

The shaking of hands was greeted by a shout of

Liverpool greets Birkenhead

From the Lord Mayor of Liverpool

And wishes all prosperity to Merseyside. Through the genius of Sir Archibald Salvidge in conceiving this tunnel Birkenhead and Liverpool will go forward to great things.

In reply, the Mayor of Birkenhead remarked,

This is one of the most momentous occasions of our lives, and I hope that all we have dreamed and wished in connection with this work will come true.

At a commemorative luncheon in the Adelphi Hotel, just up from Lime Street the Bishop of Liverpool, Sir Henry Maybury (Director-General of Roads, Ministry of Transport), Sir Max Muspratt, Sir Frederick Bowring, Col. Stott, Lt.-Col. Buckley (chairman of the Chamber of Commerce), Mr P. J. Marquis (chairman of the Liverpool Organisation), joined the other dignitaries in celebrating the momentous achievement and congratulating everyone from the committee right down to the workers at the head of the tunnel.

Sir Archibald Salvidge on behalf of the contractors was presented with a silver loving cup, a miniature replica of which was given to the two civic chiefs and the members of the Tunnel Committee. All the workmen received an illuminated certificate recording their share in the operations, and an extra day's pay.

The Mersey Tunnel Bill was, at the time a chance to help the economy of the area travel across the river and at the same time give a large unemployed community some dignity and respect in having employment security. The bill went through the house of Parliament for the ensuing session through the suspension of standing orders. It reached the House of Lords on 15th June 1925 and after a hearing of thirteen days, it finally passed. The findings were later endorsed by the committee at the House of Commons after a further eight-day hearing. The bill finally received its Royal Assent on 8th August.

Just before Christmas of 1928, it was confirmed to a shocked general public that Sir Archibald Salvidge, had passed away after months of illness. There was regret that Sir Archibald was not able to see the completion of the tunnel that he had devoted six years of his life. His successor was to be Sir Thomas White who was to become the leadership of the Liverpool Conservative party and the chairmanship of the Tunnel Committee. Under his guidance, steady progress was made with the tunnel and a contract had been agreed for the Birkenhead land tunnels. In March 1929, the fourth of the great constructional contracts tunnels was signed.

The first and only serious setback to the progress of the tunnel occurred on the 29th October, 1930. A portion of Dale Street above the excavation and close to the Police station had collapsed. This was believed to be due to the line of the tunnel crossing of the old fortifications of Cromwellian days. That collapse caused some obvious disruption to street traffic for a considerable time, but did not seriously delay operations underground. It was hoped that completion would be the summer of 1932, which would see the completion of the works. The full extent of the problem of ventilation within the tunnel was soon to rear its ugly head with the problems and financial implications that would follow.

Disclosure of the possible ventilation problems within the Mersey Tunnel were first seen in an incident in The Clifford Milburn Holland Tunnel (more commonly known as the Holland Tunnel). This is a road tunnel under the Hudson River connecting Interstate 78 on the island of Manhattan in New York City, and Route 139 in Jersey City, New Jersey on the mainland of the United States. This was seen as wakeup call as to the possible problems that would be seen in the Queensway Tunnel given its length and complexity. This caused a good deal of public anxiety and criticism when it was realised that additional funds would be needed for the tunnel. Sir Thomas White, Chairman of the committee, devised plans to raise the funds and protect the ratepayers from another financial burden. He approached the Ministry of Transport for an additional grant for the money needed to cover the new costs. His reasoning for this was that within the Mersey Tunnel Act of 1925 allowed for an increase in costs which were beyond the original estimate the Minister would consider the request.

This additional works was to allow for engineering or similar difficulties that would arise from the construction of this fledgling type of project. However, there was a sting in the tail in that the Minister would only fund a maximum of 50% from the road fund and the remainder would have to be found by the committee.

The minister noting the difficulties of the tunnels construction and the new methods and techniques that were being discovered and used during the construction, he could not allow the request. His reasons were that there were no resources at his disposal to access the additional finds, but in addition to this the state of the nation's finances were poor and the money had to be used wisely. Once this refusal was made, it became apparent that the tunnel committee had to go back to Parliament and ask for a fourth bill. This would allow the ventilation works to be completed and therefore borrow the funding. It would also request the extension of the toll charging period from 25 to 40 years in-order to keep the rate charged at a reasonably low level.

A Bill was approved by both boroughs, without a poll, in December, 1932. In the course of the session of 1933 it passed through the various Parliamentary stages without any vital amendment. According to the new scheme of finance, the estimated total cost was now given as £7,723,000 (£458,224,897.50). Deducting the Government grant of £2,500,000, (£148,331,250.00) Birkenhead's contribution of £55,000, (£3,263,287.50) and some small items, the amount to he borrowed for 80 years stood at £5,122,000, (£303,901,065.00).

For the service of that debt and all other purposes, including contributions to an equalisation fund of £4,000,000 (£237,330,000.00) to take the place of tolls at the end of 40 years, the cost of collection and management, and the creation of a renewal fund, annual revenue of £363,000 (£21,537,697.50) were estimated to be necessary. To balance the account, including revenue from tolls (in the tenth year) of £220,000 (£13,053,150.00) was assumed, while Liverpool rates were to contribute £123,000 (£7,297,897.50) and the Birkenhead rates £20,000 (£1,186,650.00).

In an article in the Liverpool Post and Mercury on July 23rd 1931, there was an article about the Tunnels and Ferries, Control of Birkenhead Services Joint Committee Decision. It states that the Mersey Tunnel Joint Committee (MTJC) took two important steps affecting the future control of the Birkenhead Ferry and Tunnel management when the tunnel opens for traffic.

As manager Mr B. H. M Hewitt (Tunnel Works Engineer) was appointed on £2,500 per annum (£148,331.25). The ferries resolved that the Birkenhead Corporation for and on behalf of the MTJC under and in accordance with such direction as may from time to time be given by the tunnel committee under section 63 of the Mersey Tunnel Act 1925. The resolution was carried by the notes of the Liverpool representatives on the committee and opposed by the whole of the Birkenhead Members who wished to remain the passenger service under sole control of the Birkenhead Corporation.

Sir Thomas White, Chairman of the Tunnel Committee, interviewed at the end of the meeting said,

"The Liverpool members were influenced to their attitude in regards to the ferries because parliament had contemplated the inclusion of the whole of the Birkenhead Ferries for 21 years as part of the traffic understanding affected by the tunnel.
"They felt," he added, "that there should be no lack of harmony or co-ordination in the cross-river services and that in the interest of the public generally. Tunnels and ferries should come under the control of one body. Some of the members felt, indeed that this principle of co-ordination should be extended to Trams, buses and Trains on both sides of the river. Further, they thought that any loss sustained by the ferries in consequence of the opening of the tunnel should not be borne by the Birkenhead Corporation and that any profits should go to reduction of the burden on the two corporations."

Alderman Clarke, of Birkenhead stated,

"The Birkenhead members opposed the transfer of the passenger services because it was the gateway to the borough. The passenger service worked closely in conjunction with the motor buses and trams which were a good paying proposition. There had been no profit for many years from the passenger service so that no financial question was involved. The whole of the ferry profits came from the goods service and by the loss of the profit Birkenhead's liability in respect of the tunnel was increased from the statutory 4d to 10d (shillings) in the pound (£0.97 – £2.43 in £58.32)."

Sir Thomas White, replying said,

"The tunnel committee did not contemplate diminishing the passenger ferry service or incommoding anybody using Birkenhead's great marine freightway. By the outside world, the change of control would be un-noticed. "The ideal," he added, "would be a federates service over and under the river, and on both sides with combined tickets and time tables."

The article went on to state, The Post understands Section 63 of the Tunnel act 1925 provides that the Birkenhead Ferry undertaking or such part of it as the Birkenhead Corporation and the Tunnels Committee may agree upon shall for 21 years from the opening of the tunnel for public traffic be worked and maintained by the Birkenhead Corporation for and on behalf of the Tunnel Committee 1925 under and in accordance with under directions as may from time to time may be given by the tunnel Committee.

The revenues of the ferries are paid to the Tunnel Committee to be responsible for the cost of wages and managing the undertaking and interest and sinking fund charges. The Committee appointed Mr H. M. Hewitt (Engineer in charge of the tunnel works) as Manager of the tunnel at £1,800 per annum (£106,798.50) with a retainer of £700 Per annum (£41,532.75) for special services which may be necessary. The appointment was for at least 12 months and if possible, 2 years. Mr Hewitt, a resident of Hoylake, has been engineer in charge of Mersey Tunnel Works under Sir Basil Mott from the beginning (5 Years ago). He has exhaustive if not unique experience in tunnel engineering. His first experience was under Mr Mott (as he was then) in connection with the Central London Railway and afterwards the City and South London railway. In 1990, he joined a geographical expedition to the Himalayan Mountains (Kashmir) and on his return, was engaged on further railway work in London.

In 1904, he went to the US for the Pennsylvania railways and was placed in charge of the tunnel under the river in New York. He remained in the US until 1912 when he went to Mexico for another tunnel. Later, he returned to the US, bored more tunnels and became a director of a firm who acted as consultants for the great (Milburn Holland Tunnel and responsible for the Queensway Ventilation System) Holland Tunnel under the Hudson River. In 1925, he returned to the UK and was placed under Sir Basil Mott in charge of the principle side of the Mersey Tunnel. Work which, once complete would rank as one of the greatest engineering feats of modern times

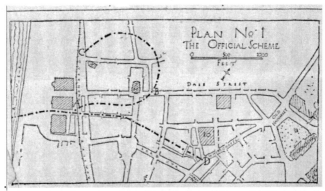

Tunnel Plan Showing the Abandoned Whitechapel Tunnel (D) (Wirral Archives)

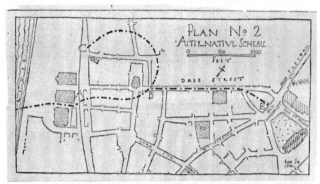
Tunnel Plan Showing the Existing Tunnel Layout (Wirral Archives)

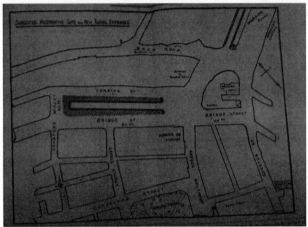
Abandoned Queensway Tunnel Entrance Proposal as Required by the Chambers of Commerce (Wirral Archives)

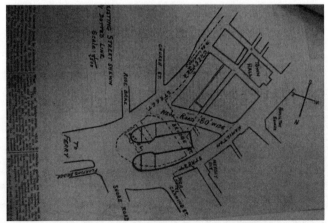
Abandoned Queensway Woodside Entrance (Wirral Archives)
(More recently occupied by Woodside Hotel
Prior to its demolition following a fire)

Tunnel Proposals circa 1923 (Wirral Archives)

Queensway (Wirral) Tunnel Proposal (Wirral Archives)

Chapter 5
Finances

Problems of Finance (1924–1925)

In May 1924, it was pointed out by Sir Archibald Salvidge that the Minister of Transport had offered to contribute not more than a third of the cost of the tunnel, after deducting £1,500,000 (£79,515,750), the estimated cost of the tramways. Inclusive of the proposed Wallasey branch, the estimated total capital outlay now stood at £7,250,000 (£384,326,125), made up as follows: Tunnel works, £5,410,000; (£286,786,805) land and expenses, £990,000; (£52,480,395) and interest during construction, management, and loss on ferries, £850,000 (£45,058,925). The maximum Government offer was £1,750,000, (£92,768,375) and it was conditional on the under-river motorway being toll-free. All this, it was calculated, would involve a rate of 5 ½ d (£1.33) in the pound over the four boroughs, and the offer was declared emphatically to be inadequate.

Opposition at this juncture seemed to gather strength. It was announced that the Dock Board was strongly antagonistic to the Wallasey branch, and it was contended that a tram service through the lower deck of the tunnel was impracticable, because only a small proportion of the ferry passengers could be carried and because the fares that would have to be charged would be prohibitive. Sir Archibald Salvidge, however, remained stubbornly optimistic. The advent of a Labour Government brought Mr Harry Gosling to the Ministry of Transport, and to hint in July with Sir Archibald stressing the urgency of finding work for the unemployed, put forward a claim to a grant of three-fourths of £5,750,000, the total cost exclusive of tramways. Mr Gosling, who had to argue the point with the new Chancellor of the Exchequer (Mr Philip Snowden), was unable to comply, but he did not close the door to further conversations.

At the end of September came the offer from Mr Gosling of £2,375,000 (£125,899,937.50) in cash, coupled with the former stipulation that the tunnel, being grant-aided, must be free of tolls. This, too, though a much-improved gesture, was declared by the Co-ordination Committee to be inadequate and unacceptable. In November, a Conservative Government was again in power, with Sir Wilfred Ashley at the Ministry of Transport, but the year 1924 closed with the tunnel scheme still in the air, or according to some of its enemy's dead and without hope of resurrection.

By the end of January, 1925, an offer on behalf of the Government had been made and accepted by the promoters as a possible basis for an application to Parliament for powers to construct a tunnel. Mr Winston Churchill was at the Treasury, and Sir Archibald Salvidge had previously put on record that but for his foresight and imagination along with freedom from red tape there would have been no Mersey Tunnel. He announced quite blankly, "That as our tunnel was to serve exactly the same purpose as a bridge, and that as a bridge, in the case of the Mersey, was not technically practicable, the objection was bureaucratic rubbish, and we received the terms we desired."

As far as a capital contribution was concerned, the previous figure of £2,375,000 (£125,899,937.50) remained unchanged, but permission was given to charge tolls for fifteen years, a concession of a capitalised value of £787,000; (£41,719,263.50) and in addition a contribution was promised towards the interest on the cost of tramways, estimated to be equal to a further grant of £400,000 (£21,204,200). The new offer therefore totalled £3,562,000, (£188,823,401) or £1,187,000 (£62,923,463.50) better than Mr Gosling's, and it was reckoned that the cost to the rates of the four boroughs would be reduced to 2 ½ d (£0.66p) in the pound.

On February 10th, the co-ordination committee resolved unanimously to proceed with the scheme, subject to the municipalities being advised that the details were practicable from the local points of view. In view of subsequent happenings, it is interesting to recall that at this juncture a good deal of uneasiness developed in Birkenhead and Wallasey in regard to the future of the ferries and the losses likely to be sustained through the absorption by the tunnel of both the vehicular and the passenger traffic.

Hostility to the scheme was becoming more and more vocal, and what appeared like a gathering storm no doubt prompted a newspaper correspondent to suggest, rather prophetically, that effort should be concentrated on the provision of a roadway for motor vehicles only, so that unanimity of public opinion might be preserved. To add to the perplexity of the supporters of the scheme, Mr Richard Holt dropped a bombshell in the shape of a protest on behalf of the Mersey Docks and Harbour Board against the proposed branch to Seacombe, which he declared would be obstructive of dock developments on the Cheshire side.

The following report is a summary of a larger document dated April 1924 outlining the costs of the proposed tunnel along with the impact on the ferries and alternative routes.

MERSEY CROSS RIVER TRAFFIC
BRIDGE OR TUNNEL
FINANCIAL REPORT
BY
MR ARTHUR COLLINS FSAA
WITH TABLES ANNEXED
20 Abingdon Street
Westminster SW1
April 1924

Page 1

To the Chairman (The Right Hon. Sir Archibald Salvidge, P.C., K.B.E.),

And Member of the Merseyside Municipal Co-ordination Committee (Special traffic Sub-Committee on Cross River Traffic)

After the initial introductions, as to the subsequent investigations, Mr Collins went on to describe the various options and areas within that investigation for the proposed tunnel.

Bridge Scheme

It was stated that although he has had to deal with both a bridge and tunnel scheme, the data collected and consolidated, relating to the bridge is only general. If the committee wish a more detailed financial breakdown of the bridge, this can be accommodated.

Tunnel Scheme

In reporting the option for the tunnel scheme, one can either present the facts relating to the existing ferry services and railway. One may consider the finance of the proposed tunnel and the possible impact of the existing cross river services.

The present and the proposed services however are almost inseparable from the financial point of view. The committee may be interested in considering their policy regarding the existing services after examination of the following.

Engineers Estimated Costs

ITEM	COST
Estimated cost of land and properties to be acquired to make room for the tunnel and its approaches	£570,000 (£30,784,521.81)
The estimated Cost of Construction	£5,410,000 (£282,182,917.53)
Parliamentary, Legal and engineering Costs	£420,000 (£22,683,331.86)
Total Costs	£6,400,000 (£345,650,771.20)
Estimated Working Expenses per annum from an Engineering Point of View	£35,000 (£1,890,277.66)

Allowance for Other Outlay

The expenditure to which the engineers have to make provision is not limited to the total sum for which a financial adviser has to provide, and although, there are other minor elements, the principle items to which attention has to be directed, in addition to engineering costs are:

- Interest on Capital during Construction
- Management and supervision during construction so far as not covered in the engineering items
- Annual Administration expenses and management during each year after completion
- The effect on the revenue of the Ferries by the loss of traffic induced to leave the Ferries by the new tunnel
- The loss of rates by the destruction of property on land to be taken over for the bridge of tunnel approaches
- Incidence of Costs

I requested the consulting engineers to assist me with the following summary calculations, namely:
1. What would the tunnel cost if it was constructed for tramway purposes alone, carrying two lines?
2. What would be the saving in the capital costs of the tunnel if it did not provide a separate segment of the tunnel for tramway purposes?

The engineers subsequently informed me of their deliberations that in their option, a tramway by itself would cost £1,500,000 (£81,011,899.50) and the saving in the cost of the present tunnel if there were no separate provision made for tramways would be £500,000 (£27,003,966.50). This would still leave a roadway taking four lanes of vehicular traffic.

Revenue Necessary from Tolls

My estimate of the annual expenditure to be covered by tolls before allocating any part to the tramways is as follows.

ITEM	COST
Maintenance Expenses (Engineers Figure)	£35,000 (£1,890,277.66)
Rates	£56,000 (£3,024,444.25)
Management, including toll takers, administration, expenses, office and establishment incidentals	£4,000 (£216,031.73)
Total without loan charges	£95,000 (£5,130,753.64)
Loan Charges as already set out	£340,000 (£18,362,697.22)
Total Annual Outgoings	£435,000 (£23,493,450.86)

The tolls required to cover these expenses and make the scheme self-supporting would be too heavy to contemplate, as it would mean making the charge to great to ensure full traffic. Without state aid, and therefore, the charge per user would appear to be prohibitive. Basing the number of potential number of users upon the best information one can obtain the result of attempting to make the scheme self-supporting by tolls may be indicated by the following computations, assuming a different volume of traffic; for instance:

The prospect of securing this traffic may be judged by comparing it with the present volume of traffic by Ferry. In 1923, the ferries carried over 745,000 vehicles and 40,000,000 passengers, on average of 2400 vehicles per weekday and 130,000 passengers per week-day. The local expectation is that the number of vehicles crossing the river would be about doubled and that fifty to seventy per cent of the passengers would cross by the tunnel instead of the ferry.

Freedom from Tolls

So far as can be ascertained, the official view is at present that it would be a condition of State Aid that the tunnels should be free of tolls, and if that view was to be maintained, having regards to the ferry losses as well as the cost of the tunnel, it scarcely seems likely that the undertaking could be made self-supporting without State Aid by tills.

State aid, therefore, appearing to be essential, coupled with a provision that the tunnel should be toll free, it is manifested that in lieu of tills the balance of expenses of the undertaking (after deduction of Government Grants) would have to be covered by rate aid. Some revenue may be expected from advertisements if the tunnel be made available, although opinion may be divided upon its desirability of taking revenue from this source.

Freedom from Tolls Calculations

As noted in the above document, there was a section for freedom of tolls even back then. We now know this is not the case due to the age and constant maintenance of the tunnels. The only way the tunnels will ever be toll free will be taking them into the national road network, which is something successive governments have refused to do. The following is an extract from that document on the toll-free argument.

Some revenue may be expected from advertisements if the tunnel be made available to advertisers, although the opinion may be divided upon the desirability of taking revenue from this source, especially as in relation to the total sum required, the revenue from advertisements could be but small. There may also be made available sites for new buildings in some places where the tunnel makes a convenient frontage for these purposes and rents may thus be secured. At present, however, no estimate of this revenue, if any, can be made.

With a toll-free arrangement, the expenses of operation as outlined would be reduced by the elimination of the allowance to be made for rates and the expenses of management in the absence of toll takers, cashiers, clerical assistance and the like. These reductions in the working expenses would amount approximately to £60,000, leaving the expenses (other than loan charges) at £35,000, this being the estimate of the Engineers to cover the maintenance of the tunnel, based on their experience of the other comparable tunnels such as Rotherhide and Blackwall.

Under the heading 'State Aid', which followed the above, the following was noted.

Conference with Sir Archibald Salvidge has had with Minister of Transport in this and the former government have led to the conclusion that as at present advised the State departments could only recommend a grant of not more than one third of the cost of the tunnel after deducting the allowance to the tramways. Such state assistance would thus, at one third, amount to £1,750,000, and this would reduce the charge for interest on capital during construction, the annual loan charges would be about £250,000, and if the tramways bore say £1,500,000 of capital, the loan charges would be £180,000 per annum, the tramways thus bearing say £70,000 a year.

Produce of Tolls at Various Rates on Possible Traffic

Class of User	Toll Per journey	Number of Users (Assumed)	Amount of Revenue
	s. d. (Shillings and Pence)	No	£
Vehicles	1s 6d (£4.05)	1,200,000	£90,000 (£4,860,713.97)
Passengers	0s 1d (£0.23)	20,000,000	£83,000 (£4,482,658.44)
			£173,000 (£9,343,372.41)
4000 vehicles per week and 64,000 passengers per week-day			
Vehicles	2s 0d (£5.40)	1,200,000	£120,000 (£6,480,951.96)
Passengers	0s 1.5d (£0.23)	20,000,000	£125,00 (£6,750,991.63)
			£245,000 (£13,231,943.59)
Vehicles	2s 6d (£6.75)	1,200,000	£150,000 (£8,101,189.95)
Passengers	0s 2d (£0.45)	20,000,000	£167,000 (£9,019,324.81)
			£317,000 (£17,120,514.76)
Vehicles	1s 0d (£2.70)	1,500,000	£75,000 (£4,050,594.98)
Passengers	0s 1d (£0.23)	30,000,000	£125,000 (£6,750,991.63)
			£200,000 (£10,801,586.66)
5000 vehicles per week and 96,000 passengers per weekday			
Vehicles	1s 6d (£4.05)	1,500,000	£112,000 (£6,048,888.50)
Passengers	0s 1.5d (£0.23)	30,000,000	£188,000 (£10,150,491.40)
			£300,000 (£16,202,379.90)
Vehicles	2s 6d (£6.75)	1,500,000	£187,000 (£10,099,483.47)
Passengers	0s 2d (£0.45)	30,000,000	£250,000 (£13,501,983.25)
			£437,000 (£23,601,466.72)

ITEM	TOTAL COST WITHOUT STATE AID Column A	NET COST WITH ONE THIRD STATE AID (EX TRAMWAY) Column B	NET COST WITH HALF STATE AID (EX TRAMWAY) Column C
Constructional Cost of works	£6,400,000 (£345,650,771.20)	£4,650,000 (£251,136,888.45)	£6,400,000 (£345,650,771.20
State aid if in Lump Sum		£1,750,000 (£94,513,882.75)	£2,500,000 (£135,019,832.50)
Capital to be locally provided	£6,400,000 (£345,650,771.20)	£4,650,000 (£251,136,888.45)	£3,900,000 (£210,630,938.70)
Interest on Capital during construction (approx.)	£850,000 (£45,906,743.05)	£600,000 (£32,404,759.80)	£500,000 (£27,003,966.50)
Total Capital Required	£7,250,000 (£391,557,514.25)	£5,250,000 (£283,541,648.25)	£4,400,000 (£237,634,905.20)
Loan Charges (when in full force)	£340,000 (£18,362,697.22)	£250,000 (£13,501,983.25)	£210,000 (£11,341,665.93)
Allocation of capital for illustration only			
To Tramways	£1,500,000 (£81,01,899.50)	£1,500,000 (£81,01,899.50)	£1,500,000 (£81,01,899.50)
To Roadways	£5,750,000 (£310,545,614.75)	£3,750,000 (£202,529,748.75)	£2,900,000 (£156,623,005.70)
Allocation of Loan Charges on basis if above			
To Tramways	£70,000 (£3,780,555.31)	£70,000 (£3,780,555.31)	£70,000 (£3,780,555.31)
To Roadways	£270,000 (£14,582,141.91)	£180,000 (£9,721,427.94	£140,000 (£7,561,110.62)
Total Loan Charges	£340,000 (£18,362,697.22)	£250,000 (£13,501,983.25)	£210,000 (£11,341,665.93)

Maintenance expenses £35,000 per annum			
To Tramways	£5000 (£270,039.66)	£5000 (£270,039.66)	£5000 (£270,039.66)
To Roadways	£30,000 (£1,620,237.99)	£30,00 (£1,620,237.99)	£30,000 (£1,620,237.99)
Total approx., charge on Rates after deducting 50% grant on maintenance of roads parts of tunnel (Column B and c)	£300,000 (£16,202,379.90)	£195,000 (£10,531,546.94)	£155,000 (£8,371,299.62)
Rate in £ of aid required taken to say £35,000 per 1d rate as from 1930	8 1/2d (£1.80)	5 1/2d (£1.13)	4 1/2d (£0.90)

MERSEY CROSS RIVER TRAFFIC
(TUNNEL SCHEME)

Schedule of calculations of Annual Loan Charges for the first 30 years assuming 80 years loans for the tunnel, with sinking fund accumulating at 3.5%

Year	Total capital expenditure	Interest to be capitalized during construction	Total capital expenditure after capitalizing interest	Interest on col 4	Rate of interest	Sinking fund at 3.5% (80 years)	Total annual charges after capitalization
	£	£	£	£	£	£	£
1	£6,400,000 (£345,650,771.20)	£838,000 (£45,258,647.85)	£7,238,006 (£390,909,743.10)	£289,520 (£15,636,376.76)	4%	£17,262 (£932,284.94)	£306,782 (£16,568,661.70)
2	£6,400,000 (£345,650,771.20)	£838,000 (£45,258,647.85)	£7,238,006 (£390,909,743.10)	£289,520 (£15,636,376.76)	4%	£17,262 (£932,284.94)	£306,782 (£16,568,661.70)
3	£6,400,000 (£345,650,771.20)	£838,000 (£45,258,647.85)	£7,238,006 (£390,909,743.10)	£325,710 (£17,590,923.86)	4.5%	£17,262 (£932,284.94)	£342,972 (£18,523,208.80)
4	£6,400,000 (£345,650,771.20)	£838,000 (£45,258,647.85)	£7,238,006 (£390,909,743.10)	£325,710 (£17,590,923.86)	4.5%	£17,262 (£932,284.94)	£342,972 (£18,523,208.80)
5	£6,400,000 (£345,650,771.20)	£838,000 (£45,258,647.85)	£7,238,006 (£390,909,743.10)	£325,710 (£17,590,923.86)	4.5%	£17,262 (£932,284.94)	£342,972 (£18,523,208.80)
6	£6,400,000 (£345,650,771.20)	£838,000 (£45,258,647.85)	£7,238,006 (£390,909,743.10)	£325,710 (£17,590,923.86)	4.5%	£17,262 (£932,284.94)	£342,972 (£18,523,208.80)
7	£6,400,000 (£345,650,771.20)	£838,000 (£45,258,647.85)	£7,238,006 (£390,909,743.10)	£325,710 (£17,590,923.86)	4.5%	£17,262 (£932,284.94)	£342,972 (£18,523,208.80)
8	£6,400,000 (£345,650,771.20)	£838,000 (£45,258,647.85)	£7,238,006 (£390,909,743.10)	£325,710 (£17,590,923.86)	4.5%	£17,262 (£932,284.94)	£342,972 (£18,523,208.80)
9	£6,400,000 (£345,650,771.20)	£838,000 (£45,258,647.85)	£7,238,006 (£390,909,743.10)	£325,710 (£17,590,923.86)	4.5%	£17,262 (£932,284.94)	£342,972 (£18,523,208.80)

	Total capital expenditure	Interest to be capitalized during construction	Total capital expenditure after capitalizing interest	Interest on col 4	Rate of interest	Sinking fund at 3.5% (80 years)	Total annual charges after capitalization
10	£6,400,000 (£345,650,771.20)	£838,000 (£45,258,647.85)	£7,238,006 (£390,909,743.10)	£325,710 (£17,590,923.86)	4.5%	£17,262 (£932,284.94)	£342,972 (£18,523,208.80)
11	£6,400,000 (£345,650,771.20)	£838,000 (£45,258,647.85)	£7,238,006 (£390,909,743.10)	£325,710 (£17,590,923.86)	4.5%	£17,262 (£932,284.94)	£342,972 (£18,523,208.80)
12	£6,400,000 (£345,650,771.20)	£838,000 (£45,258,647.85)	£7,238,006 (£390,909,743.10)	£325,710 (£17,590,923.86)	4.5%	£17,262 (£932,284.94)	£342,972 (£18,523,208.80)
13	£6,400,000 (£345,650,771.20)	£838,000 (£45,258,647.85)	£7,238,006 (£390,909,743.10)	£325,710 (£17,590,923.86)	4.5%	£17,262 (£932,284.94)	£342,972 (£18,523,208.80)
14	£6,400,000 (£345,650,771.20)	£838,000 (£45,258,647.85)	£7,238,006 (£390,909,743.10)	£325,710 (£17,590,923.86)	4.5%	£17,262 (£932,284.94)	£342,972 (£18,523,208.80)
Year	Total capital expenditure	Interest to be capitalized during construction	Total capital expenditure after capitalizing interest	Interest on col 4	Rate of interest	Sinking fund at 3.5% (80 years)	Total annual charges after capitalization
15	£6,400,000 (£345,650,771.20)	£838,000 (£45,258,647.85)	£7,238,006 (£390,909,743.10)	£325,710 (£17,590,923.86)	4.5%	£17,262 (£932,284.94)	£342,972 (£18,523,208.80)
16	£6,400,000 (£345,650,771.20)	£838,000 (£45,258,647.85)	£7,238,006 (£390,909,743.10)	£325,710 (£17,590,923.86)	4.5%	£17,262 (£932,284.94)	£342,972 (£18,523,208.80)
17	£6,400,000 (£345,650,771.20)	£838,000 (£45,258,647.85)	£7,238,006 (£390,909,743.10)	£325,710 (£17,590,923.86)	4.5%	£17,262 (£932,284.94)	£342,972 (£18,523,208.80)
18	£6,400,000 (£345,650,771.20)	£838,000 (£45,258,647.85)	£7,238,006 (£390,909,743.10)	£307,615 (£16,613,650.31)	4.5%	£17,262 (£932,284.94)	£342,877 (£18,518,078.04)
19	£6,400,000 (£345,650,771.20)	£838,000 (£45,258,647.85)	£7,238,006 (£390,909,743.10)	£307,615 (£16,613,650.31)	4.5%	£17,262 (£932,284.94)	£342,877 (£18,518,078.04)
20	£6,400,000 (£345,650,771.20)	£838,000 (£45,258,647.85)	£7,238,006 (£390,909,743.10)	£307,615 (£16,613,650.31)	4.5%	£17,262 (£932,284.94)	£342,877 (£18,518,078.04)

21	£6,400,000 (£345,650,771.20)	£838,000 (£45,258,647.85)	£7,238,006 (£390,909,743.10)	£307,615 (£16,613,650.31)	4.5%	£17,262 (£932,284.94)	£342,877 (£18,518,078.04)
22	£6,400,000 (£345,650,771.20)	£838,000 (£45,258,647.85)	£7,238,006 (£390,909,743.10)	£307,615 (£16,613,650.31)	4.5%	£17,262 (£932,284.94)	£342,877 (£18,518,078.04)
23	£6,400,000 (£345,650,771.20)	£838,000 (£45,258,647.85)	£7,238,006 (£390,909,743.10)	£289,520 (£15,636,376.76)	4%	£17,262 (£932,284.94)	£306,782 (£16,568,661.70)
24	£6,400,000 (£345,650,771.20)	£838,000 (£45,258,647.85)	£7,238,006 (£390,909,743.10)	£289,520 (£15,636,376.76)	4%	£17,262 (£932,284.94)	£306,782 (£16,568,661.70)
25	£6,400,000 (£345,650,771.20)	£838,000 (£45,258,647.85)	£7,238,006 (£390,909,743.10)	£289,520 (£15,636,376.76)	4%	£17,262 (£932,284.94)	£306,782 (£16,568,661.70)
26	£6,400,000 (£345,650,771.20)	£838,000 (£45,258,647.85)	£7,238,006 (£390,909,743.10)	£289,520 (£15,636,376.76)	4%	£17,262 (£932,284.94)	£306,782 (£16,568,661.70)
27	£6,400,000 (£345,650,771.20)	£838,000 (£45,258,647.85)	£7,238,006 (£390,909,743.10)	£289,520 (£15,636,376.76)	4%	£17,262 (£932,284.94)	£306,782 (£16,568,661.70)
28	£6,400,000 (£345,650,771.20)	£838,000 (£45,258,647.85)	£7,238,006 (£390,909,743.10)	£289,520 (£15,636,376.76)	4%	£17,262 (£932,284.94)	£306,782 (£16,568,661.70)
29	£6,400,000 (£345,650,771.20)	£838,000 (£45,258,647.85)	£7,238,006 (£390,909,743.10)	£289,520 (£15,636,376.76)	4%	£17,262 (£932,284.94)	£306,782 (£16,568,661.70)
30	£6,400,000 (£345,650,771.20)	£838,000 (£45,258,647.85)	£7,238,006 (£390,909,743.10)	£289,520 (£15,636,376.76)	4%	£17,262 (£932,284.94)	£306,782 (£16,568,661.70)

Two Stage Settlements in (1927–1928)

Despite this result, none of the parties to the dispute was satisfied with the prospect. In the course of the following two months' discussions took place, in which Lord Birkenhead, then the Secretary of State for India, acted privately as mediator, and as a result of his diplomacy, it was announced in mid-April 1927, that agreement had been reached. According to the new plan, Birkenhead was to have two entrances. The main Woodside entrance was to be set back to the corner of Bridge Street and Chester Street, and a branch, 27 ft wide, was to be made to Rendel Street, at an extra cost of £222,000 (£12,046,166).

In lieu of paying one-half of this sum, as was suggested, the Ministry of Transport sanctioned a further extension of the toll period to 25 years. The Birkenhead Corporation paid a quarter of the amount, which was £55,000 (£3,011,541.50), and the other quarter became a liability of the Tunnel Committee. A 'treaty of peace' was formally signed by the Tunnel Committee, and the happy ending, as it was thought to be, was later celebrated by a Lord Mayoral banquet, with the peacemaker, Lord Birkenhead, as the guest of honour.

The amending bill, now sponsored by both Corporations, had a smooth passage through Parliament during the 1927 session, and received the Royal Assent on July 27th. But, though it embodied the agreement arrived at, it by no means, as it turned out, represented finality as regards the tunnel plans. A still bigger departure from the original recommendations of the engineers was to come.

Opinion in Liverpool that the Old Haymarket possessed substantial advantages over Whitechapel as the site of the main city entrance had been steadily gathering force since the passing of the first Tunnel Act. But, though it was freely voiced in the City Council and the Press, the Tunnel Committee held on its course, apparently unaffected by the representations, and a beginning was actually made with the excavation from underneath Brunswick Street in the direction of Whitechapel.

In Birkenhead, too, there was still much uneasiness. The signing of the 'peace treaty', which gave the borough a dock entrance, precluded another public agitation for a further change, but nobody liked the Bridge Street proposal. Now that the dock traffic was to be separately accommodated, Woodside had lost its attractiveness, and the view was held almost unanimously that both the town and the regions beyond would be best served by an entrance in the vicinity of the Haymarket.

Ever a lover of drama in public affairs, Sir Archibald Salvidge caused a sensation on November 4th by submitting to the Tunnel Committee new and unheralded proposals, formulated, it was stated, after a close study of the problem by hint in consultation with the engineers (Messrs. Mott & Brodie), the valuer (Mr F. J. Kirby), and the clerk (Mr 'Walter Moon).

Under the new scheme the line to Whitechapel, Liverpool, was to be abandoned, and instead a new line, parting with the New Quay loop at a point underneath the State Insurance offices, was to be constructed beneath Dale Street to its junction with Manchester Street. There it would enter the triangle of land on the right owned by the Corporation and earmarked for new municipal offices, which would provide the necessary open cutting and permit of an exit to the Old Haymarket, facing St. John's Gardens and St George's Hall.

On the Birkenhead side, the main tunnel line would avoid the curve backward to Woodside and proceed beneath Hamilton Square and Albion Street to Market Street South, near the Haymarket, a point giving easy access to both of the main roads to Chester and the Dee Bridge. Involved in the adoption of this site was the demolition of the Carnegie Free Library and the building in Borough Road of a new library, to be opened by the King in July 1934.

It was officially stated that these alterations would not mean any extra cost owing partly to a saving in expenditure on property and easements, and partly to the discovery that some of the cementation and other work originally estimated for could be dispensed with. At both ends of the main tunnel, the gradients would be improved—Liverpool from 1 in 20 to 1 in 30, and at Birkenhead from 1 in 20 to 1 in 28. In the case of Liverpool, an overriding consideration was the saving of about £11,300,000 (£618,734,890) on a new wide road to connect Whitechapel with Lime Street. While the projected Everton tunnel road, from Byrom Street to the new East Lancashire road, held to be necessary in any case, constituted an additional argument endorsed by public opinion as well as by the two Councils in favour of the Old Haymarket.

The revised plans were considered and Mersey Tunnel Act, No. 3, which received the Royal approved by the committee unanimously, and the decision received Assent on April 26th, 1928.

PRIVATE AND CONFIDENTIAL

MERSEY TUNNEL JOINT COMMITTEE
COMPARATIVE STATEMENT OF ESTIMATES AND EXPENDITURE
YEAR ENDING 31ST MARCH 1932

	Estimate 1931-1932	Expenditure	Balance of Estimate Unexpended
CAPITAL ACCOUNT			
Construction of Works	£1,239,000 (£73,512,967.50)	£314,020 (£54,881,375.85)	£924,980 (£54,881,375.85)
Land and easements	£170,500 (£10,116,191.25)	£47,300 (£2,806,427.25)	£123,200 (£7,309,764.00)
Interest on Loans for purchase of land and construction of works	£126,000 (£7,475,895.00)	£54,033 (£3,205,912.97)	£71,967 (£4,269,982.03)
Legal and Other expenses	£500 (£29,666.25)		£500 (£29,666.25)
TOTALS	£1,536,000 (£91,134,720.00)	£415,353 (£24,643,931.87)	£1,120,647 (£66,490,788.13)

	Estimate 1931-1932	Expenditure	Balance of Estimate Unexpended
REVENUE ACCOUNT			
Expenses of promotion of Mersey Tunnel Acts			
Instalment of Loans	£847 (£50,254.63)		£847 (£50,254.63)
Interest on Loans	£88 (£5,221.26)	£44 (£50,254.63)	£44 (£5,221.26)
Interest on Loans to liquidate capitalised interest	£8,771 (£520,405.36)		£8,771 (£520,405.36)
Expenses re Loans	£2,750 (£163,164.38)	£2,008 (£119,139.66)	£742 (£44,024.72)
Rent of Sites	£500 (£29,666.25)	£260 (£15,426.45)	£240 (£14,239.80)
Administration Expenses	£1,640 (£97,305.30)	£826 (£49,008.65)	£814 (£48,296.66)
Miscellaneous Expenses	£40 (£2,373.30)	£40 (£2,373.30)	
TOTALS	£14,636 (£868,390.47)	£3,178 (£188,558.69)	£11,458 (£679,831.79)

(Signed) W. E. LEGH-SMITH
Treasurer

09 October 1931
Letter from Town Clerk at Liverpool Municipal Building dated 3rd November 1932

My Dear Sir,
MERSEY TUNNEL EXCESS EXPENDITURE
I enclose herewith a copy of a report of the Treasurer to the Joint Committee in regard to the estimated additional expenditure in connection with the Mersey Tunnel, which will be submitted to the Joint Committee at the Special Meeting to be held at 3.30pm to-morrow, the 4th instant.
You will observe that the Report is marked 'Private and Confidential.'
Yours faithfully
Clerk

PRIVATE AND CONFIDENTIAL

MERSEY TUNNEL JOINT COMMITTEE
The treasurer begs to report that he has received from the Engineer, the Architect, and the Valuer, estimates of the ultimate cost of the tunnel.
For the purpose of comparison, the estimate prepared in February 1932, is given below, together with the revised present estimate.

(a) ENGINEERING WORKS

Item	Estimate of February 1932 £	Estimate of November 1932 £	Difference £
Total of Construction Contracts Nos 1, 2, 3, 4 and 4a	4,300,580 (£255,164,162.85)	4,364,750 (£258,971,529.38)	+64,170 (£3,807,366.53)
Ventilation Works below ground level	301,170 (£17,869,169.03)	249,000 (£14,773,792.50)	-52,170 (£3,095,376.53)
Fans, motor equipment, lighting, pumps, fire appliances, rolling stock, office equipment, etc.:	327,294 (£19,419,171.26)	249,500 (£14,803,458.75)	-77,794 (£4,615,712.51)
Street works (Plazas):	150,000 (£8,899,875.00)	169,000 (£10,027,192.50)	+19,000 (£1,127,317.50)
Claims for compensation, loss of water, etc.:	115,000 (£6,823,237.50)	115,000 (£6,823,237.50)	
Foundations for Ventilation buildings	54,000 (£3,203,955.00)	107,000 (£6,348,577.50)	+53,000 (£3,144,622.50)
Share of Costs of repair of collapsed quay wall	5,750 (£341,161.88)	5,750 (£341,161.88)	
Initial Ventilation, lighting, watching, pumping, etc., until opening of Tunnel		78,000 (£4,627,935.00)	+78,000 (£4,627,935.00)
SUB TOTALS	5,253,794 (£311,720,732.51)	5,338,000 (£316,716,885.00)	+84,206 (£4,996,152.50)
Contingencies on remaining records	29,250 (£1,735,475.63)	129,000 (£7,653,892.50)	+99,750 (£5,918,416.88)
TOTALS	5,283,044 (£313,456,208.13)	5,467,000 (£324,370,777.50)	+183,956 (£10,914,569.37)

(b) VENTILATION BUILDING

Item	Estimate of February 1932 £	Estimate of November 1932 £	Difference £
Ventilation and control station and operating offices at George's Dock, Ventilation Building at North John Street, New quay, Woodside, Sidney Street and Taylor Street:	508,000 (£30,140,910.00)	519,190 (£30,804,840.68)	+11,190 (£663,930.68)

Concrete Fan Casings and other engineering items		68,155 (£4,043,806.54)	+68,155 (£4,043,806.54)
Garage and workshop accommodation and forecourts at George's Dock not yet authorised		33,415 (£1,982,595.49)	+33,415 (£1,982,595.49)
Total estimate of building works	508,000 (£30,140,910.00)	620,760 (£36,831,242.70)	+112,760 (£6,690,332.70)
Architects fees	25,000 (£1,483,312.50)	33,540 (£1,990,012.05)	+8,540 (£506,699.55)
Contingencies – it is recommended that the contingency item be included so that the Parliamentary Bill may be include a sum which may cover all eventualities. In this contingency are included fees for extra work, the amount of which it is not yet possible to ascertain.		68,500 (£4,064,276.25)	+68,500 (£4,064,276.25)
Estimate of total provision in respect of building works and Architects fees	533,000 (£31,624,222.50)	722,800 (£42,885,531.00)	+189,800 (£11,261,308.50)

(c) LAND

Item	Estimate of February 1932 £	Estimate of November 1932 £	Difference £
Land	733,552 (£43,521,694.07)	848,000 (£50,313,960.00)	+94,448 (£5,603,835.96)

(d) SUNDRY COSTS

Item	Estimate of February 1932 £	Estimate of November 1932 £	Difference £
Legal, Surveying Costs etc.	81,448 (£4,832,513.46)	40,000 (£2,373,300.00)	-21,448 (£2,373,300.00)

SUMMARY
LANDS, BUILDINGS AND WORKS

Item	Estimate of February 1932 £	Estimate of November 1932 £	Difference £
Engineering Costs	5,283,044 (£313,456,208.13)	5,467,000 (£324,370,777.50)	+ 183,956 (£10,914,569.37)
Ventilation Buildings	533,000 (£31,624,222.50)	722,800 (£42,885,531.00)	+ 189,800 (£11,261,308.50)
Land	753,552 (£44,710,124.04)	848,000 (£50,313,960.00)	+ 94,448 (£5,603,835.96)
Sundry Costs	81,448 (£4,832,513.46)	40,000 (£2,373,300.00)	-21,448 (£1,272,563.46)
	6,631,044 (£393,436,418.13)	7,077,800 (£419,943,568.50)	+446,756 (£26,507,150.37)

The foregoing does not include either capitalised interest or promotion expenses
BORROWING POWERS

IN RESPECT OF LANDS, BUILDINGS AND WORKS

The additional amount of borrowing powers required is arrived at as under:

	£	£
Total Estimated Cost		7,077,000 (£441,253,073.10)
Deduct Government Grant	2,500,000 (£155,875,750.00)	
Amount payable to Birkenhead, being ¼ of £220,000	55,500 (£3,460,441.65)	2,555,500 (£159,336,191.65)
Deduct borrowing powers already authorised – Under Mersey Tunnel Act 1925	2,500,000 (£155,875,750.00)	
Under sanction of Ministry of Health dated 17th March, 1932:	500,000 (£31,175,150.00)	3,000,000 (£187,050,900.00)
Additional borrowing powers required in respect of Lands, Buildings and Works		£1,511,500 (£94,242,478.45)

No additional borrowing powers are required in respect of Capitalised Interest, as the Act of 1925 authorises the borrowings of 'the amount necessary' under this heading. It is estimated that the interest thus capitalised will amount to £600,000 (£37,410,180.00) in the event of the tunnel being opened for traffic by March 1934 or to £750,000 (£46,762,725.00) if not opened before March 1935.

Provision is made in the draft Bill for the borrowing of the promotion expenses in connection therewith.

(Signed) W. E. LEGH-SMITH
Treasurer
3rd November 1932

Contracts

The signing of a contract to undertake such a bold engineering project was in its self to be bold. The committee had to ensure that the chosen contractors were capable of sustaining the work and at a good standard. They had to ensure that the four-lane roadway and all associated services were completely safe to use for both vehicles and passengers alike. The tunnel had to have an absolute assurance that it would be water tight (allowing for a small seepage within acceptable limits) and in view of its length, have adequate ventilation. Finally, the lighting must be to a level that would be sufficient in allowing an almost daylight feel for the users whilst in the tunnel.

Contract No. I.—Sinking working shafts and driving pilot headings, Edmund Nuttall, Sons & Co, Ltd, Trafford Park, Manchester.

Contract No. 2.—Enlargement of pilot headings to full-sized tunnel and lining with cast-iron, section under river between George's Dock, Liverpool, and Morpeth Branch Dock, Birkenhead, Edmund Nuttall, Sons & Co, Ltd, Trafford Park, Manchester.

During the period when Contract No. 2, for the construction of the full-size tunnel under the river, was being fulfilled, steps were taken to obtain tenders for the construction of the approach tunnels on both the Liverpool and Birkenhead sides of the river

Contract No. 3.—Enlargement of pilot headings to full-sized tunnels in Birkenhead, including lining, concreting roadway and interior finish, Sir Robert McAlpine & Sons, Ltd, London.

Full specifications were accordingly issued for Contract No. 3, for the construction of the tunnel between the main entrance at Birkenhead and the shaft at Morpeth Dock, and also for the smaller tunnel to the exit at Rendel Street, which has been arranged for the convenience of goods traffic going to and from the Birkenhead docks.

Contract No. 4.—Enlargement of pilot headings to full-sized tunnels in Liverpool, including lining, concreting roadway and interior finish, Edmund Nuttall, Sons & Co, Ltd, Trafford Park, Manchester.

Contract No. 4 covered the main tunnel from the Old Haymarket to the George's Dock shaft, and also the tunnel leading to the New Quay entrance.

These contracts involved, in addition to the tunnelling, the construction of junction chambers at the points where the main and the branch tunnels met. These chambers were naturally of greater dimensions than the main tunnel itself, which was over 70 feet wide and 55 feet long. Each chamber was constructed with an arched roof.

The work involved in the tunnel construction would initially involve the construction of a shaft 200 feet (60.96 metres) deep and 21 ft 21 inches (6.934 metres) wide on each side of the river. Work was to begin with the construction of two pilot holes (one upper and one lower) 15 ft (4.57 metres) wide by 12 ft (3.65 metres) high and connecting the two shafts. Included in this was the construction of 7 ft (2.13 metres) wide drainage from the Liverpool side of the tunnel and a number of strategically placed bore holes connecting the lower tunnel to the pilot holes. Construction shafts would connect the upper and lower tunnels along the tunnel.

Edmund Nuttall was mindful of the problems that had arisen forty years earlier during the construction of the Merseyrail Tunnel by Sir Basil Mott. The experience gained during this construction project, namely the ground conditions and the river bed-tunnelling problems would prove to be invaluable during the road tunnel works. The experience gained by Sir Basil Mott during the rail tunnel had been used by others and had been enhanced by forty year of development and experience now being used on the road tunnel.

Chapter 6
Purchase of Property

The purpose of acquiring a property for a large development such as the Mersey Tunnel has taken many forms throughout history. The Act of Parliament in 1925 set the initial legal standing for the process of constructing the Mersey Tunnel. The final changes in entrances needed a good deal of additional expenditure on land and property.

The Acquisitions at Birkenhead Included The:
1. Carnegie Free Library
2. Ellis and Powell's timber yard
3. Various office buildings
4. Many shops and licenced premises

It also included the reinstatement of:

Licenced Houses
1. Motor garage in Argyle Street
2. Large steel and iron business in Cleveland Street, which was reinstated on a new site
3. A school
4. Weigh-bridge belonging to the Corporation of Birkenhead

Rutherford's shipbuilding works in Bridge Street, founded in 1853 and in 1914 Shipbuilders and Repairers, Engineers, Boilermakers, Steam Yacht and Boat Builders, Ship and Yacht Outfitters.

Specialities: Light Draft Steam and Motor Vessels of all kinds, all descriptions of small craft, ship repairs, and in 1921, it became a private company.

Carnegie Free Library was chosen as the site for the new tunnel and was replaced by the Birkenhead Central Library. The Library was opened immediately after the tunnel by the King, and is still in use today.

The surveying and acquisition of the property need for the tunnel was given to Mr V J Kirby and this was proving to be a difficult task. The Rent and Mortgage Restriction Act had inadvertently caused a number of problems with the process. Within the Act it allowed for a tenant who was paying a statutory rent could not be turned out unless suitable alternative accommodation could be offered. As such accommodation had to be convenient to allow the person to carry out their business in the heart of Birkenhead, from a suitable premise, the task was almost impossible.

There were many such tenants in Birkenhead and one in particular was that of a lodging-house in Albion Street, one half was rent-restricted while the other was not. The contractor's work-men early one morning proceeded to take the roof off what was unfortunately the occupied and rent-restricted portion, and were met with a greeting that was non-to pleasant from the occupier, that can well be imagined.

The acquisition of the Feathers Hotel in Chester Street was noteworthy because the owners, owing to its purchase for street improvements, obtained a 'special' removal of the licence to new premises in Borough Road. This step undoubtedly saved the tunnel committee a large sum of money, as only the unlicenced value of the property had to be bought.

The story of the acquisitions in Liverpool was to be a different story as little of the area for the tunnel was not covered by the 1925 act. The change to the Old Haymarket involved the purchase of an island site bounded by Dale Street, Manchester Street and the Old Haymarket, which belonged to the Corporation of Liverpool, and other freehold owners.

However, this did not mean that the land would be an easy acquisition and or cheap as the Corporation had to be seen to get a true value for the land. This would make the process more economical and negated the prospect of elongated discussions as to the final purchase price of the property or rehousing of the tenants/owners.

The total expenditure on property was found to be substantially within the original estimate of 1925. This is despite the fact that large sites for the additional six ventilating stations, including that at the George's Dock, which was to be the main tunnel headquarters, were acquired.

During the digging of the tunnel, it was made clear that the ground conditions were to be taken into consideration and so far, as possible the works were not to affect the property at ground level. This was partially covered by the cementation process in stabilising the ground within the tunnel, thus avoiding a collapse and subsequent problems above ground.

To enable this process to work correctly and avoid any collapsing of the tunnel, the contractors had to work closely. This involved the closest co-operation between Nuttall and the Cementation Co up to the point when it was possible to dispense with the process. In the first instance, the cementation contractor would apply the process to the rock formation and allow it to set. Nuttall would then send in the excavators and remove the rock. This process would be completed time after time until stable rock formations were reached.

To allow the initial processes of the tunnel to be started, two shafts (one on each side of the river) were constructed up to the level of the river. Upon reaching this level, a process of silicatisation (impregnated silicate to harden make them less sensitive to frost.) and cementation treatments were added to the shaft walls. This allowed the shafts to be sunk to the depth below the river and allow the tunnel process to begin. The depths of this additional section of tunnel were 195 (59.43 metres) on the Birkenhead side and 100 feet (30.48 metres) on the Liverpool side.

The shafts were set on the tunnel line, and the first operations included the driving of adits (a horizontal passage leading into a mine for the purposes of access or drainage) to connect with the tunnel line. The cementation of the shafts was supplemented by horizontal cementation to enable the ground, pilot holes and connecting adits to be secure. A small tunnel approximately 12 x 15 feet (3.65 x 4.57 metres) was started and placed on a specific heading along the correct tunnel line to initiate the pilot tunnels.

Radiating holes were drilled in series from the heading face, the inclination and length was calculated to cover the full section of 46 feet (14.0 metres) diameter tunnel (the full diameter of the finished tunnel) over a length of 150 feet (45.72 metres). The drilling of holes at small angles from horizontal, deflection occurs which varies with the ground drilled through, but after a little experience it was possible to calculate this deflection and to make the necessary allowances in the machine settings to ensure that the holes terminated at the desired point.

The upper heading was then driven a distance of 135 feet (41.15 metres), leaving a plug of cemented ground to form the bulkhead or dam against the water contained in the untreated ground ahead, and from the protection of this plug a further length of cementation was operated. Cementation operations alternated between the top and bottom headings in the section between the shafts. In both cases, however, the full section of 46 feet (14.02 metres) diameter tunnel was treated, the only difference being that the angle of the holes had to be altered. It should be noted that after the cemented plug or bulkhead had been penetrated, boring was in virgin ground, i.e., ground untreated by cementation and containing large quantities of water. To prevent this water entering the workings and thus throwing an excessive load on the pumping plant, control apparatus was employed during boring and prior to sealing by cementation.

In the Liverpool and Birkenhead approach tunnels, the pilot heading was in the upper part of the tunnel. It was intended that cementation should be applied from the face of this heading and over the water-bearing sections.

As the work proceeded, it was found that the inflow of water was not sufficient to warrant the continuation of the cementation treatment, and it was, therefore, stopped. At the time when the process was discontinued, there had been the following areas covered by the process, in Liverpool, 850 feet (259.08 metres) of rock from the shaft towards the river, or 160 feet (48.77 metres) beyond the quay wall, and 450 feet (137.16 metres) from the shaft landward under Liverpool. On the Birkenhead side the treatment had reached 1,170 feet (356.61 metres) from the shaft towards the river or 410 feet (124.97 metres) beyond the quay wall.

Old Bakery Camberdown and Hamilton Street c1929 (Birkenhead Library)

Library in Market Place South being Demolished c1929 (Birkenhead Library)

Market Street – Hamilton Street Demolition March 1930s (Birkenhead Library)

Cross Street Demolition c1929 (Birkenhead Library)

Old Feathers Hotel in Ruins (Birkenhead Library)

The Carnegie Central Library opened in 1909 and subsequently demolished in 1929 to make way for the Mersey Tunnel (Queensway) (Birkenhead library)

King Edward VII Clock (Grade 2) At Central Station, now moved to the middle of the Central Station Roundabout due to the Queensway Tunnel (Birkenhead Library)

King Edward VII Clock (Grade 2) Today

Chapter 7
Full Scale Experiment

With the exception of the initial requirement for a tunnel, the geotechnical and other surveys will establish if the ground is capable of sustaining the structures and its foundations. It will also establish any requirements for a Geotechnical investigation which are set up to acquire the data that will facilitate successful design, installation and the operational integrity of the structure. To this end, site specific information is required on:

1. Soil type and variability
2. Strength, deformation and consolidation characteristics
3. Topography of the land to which the tunnel will pass
4. The influence of cyclic loading
5. Scour potential
6. Land instability
7. Earthquake susceptibility
8. Presence of shallow gas or gas hydrates

Now the basic information has been acquired an informed choice can then be made of machinery and methods for excavation. This information will also show any possible locations of fault lines, soft ground or dangerous rock formations. This will greatly reduce the risk of encountering unforeseen ground conditions. The route of the tunnel can now be made with reasonable certainty. There is also the possibility of bad ground being found during the tunnelling process.

On a large diameter tunnel, such as the Queensway, a pilot tunnel, or drift, may be driven ahead of the main drive. This smaller diameter tunnel will be easier to support should unexpected conditions be met, and will be incorporated in the final tunnel. Alternatively, horizontal boreholes may sometimes be drilled ahead of the advancing tunnel face.

The method of tunnel construction depends on the geotechnical information and such factors as the length and diameter of the tunnel, the depth of the tunnel, the logistics of supporting the tunnel excavation, the final use and shape of the tunnel and appropriate risk management.

A shaft is occasionally necessary for a tunnel project. They are usually circular and go straight down until they reach the level at which the tunnel is going to be built. A shaft normally has concrete walls and is built to be permanent. Once they are built, Tunnel Boring Machines are lowered to the bottom and excavation can start. Shafts are the main entrance in and out of the tunnel until the project is completed. If a tunnel is going to be long, multiple shafts at various locations will be bored so that entrance into the tunnel is closer to the unexcavated area.

Stand-up time is the amount of time a tunnel will support itself without any added structures. Knowing this time allows the engineers to determine how much can be excavated before support is needed. The longer the stand-up time is the faster the excavating will go. Generally, certain configurations of rock and clay will have the greatest stand-up time, and sand and fine soils will have a much lower stand-up time.

Groundwater control is very important in tunnel construction. If there is water leaking into the tunnel, stand-up time will be greatly decreased. If there is water leaking into the shaft, it will become unstable and will not be safe to work in. To stop this from happening there are a few common methods in use and one of the most effective is ground freezing. To do this pipes are inserted into the ground surrounding the shaft and they are then cooled until they become semi frozen. This freezes the ground around each pipe until the whole shaft is surrounded by frozen soil, keeping water out. Pipes are installed into the ground to simply pump the water out. This works for tunnels and shafts.

Long before the pilot tunnels were begun, the engineers for the Queensway Tunnel started work on a full-scale trial to define the tunnel size. This is to allow them to make the necessary calculations on stress, tension and compression of the materials along with the surrounding strata. Given that no one had constructed a tunnel of this size before, they were in completely uncharted territory.

They decided on a trial tunnel 300 feet (91.44 metres) long and 48 feet 3 inches (14.70 metres) in diameter between the Birkenhead shaft and the river wall. The task of building the trial tunnel was given to Nuttall as an addendum to their initial contract, Contract No 1.

One of the major conclusions of this experiment was that the tunnel would be excavated in two halves. The top roadway section would be excavated and lined with a continuous ring of iron segments, before the bottom half was excavated. This would allow the rock above to be fully supported and give the workers a safer environment to work in as well as allowing such a large diameter tunnel to be constructed. There were a few important considerations to be taken for the iron segments, especially as they would be susceptible to rusting in a salt water environment.

The spacing and size of the segments would have to ensure water tightness as well as be waterproof. The size would have to be manageable but not to small so as to add additional joints to the tunnel lining. The more joints, the more likely it would be that one would fail and water would seep into the tunnel. The requirement would be for the minimum number of joints with a manageable section of iron. The contact surface between the tunnel section and the rock would also have to be carefully considered. Once the section was in place it was there for life so a suitable method of protection had to be found.

The arch of rock in the upper half of the circle would be supported by the cast-iron lining whilst the lower half of the lining was being completed, with the top half of the lining being supported meanwhile by the underlying rock. When that had been done, the gradual withdrawal of the rock support to the excavation and the lining of the lower half proceeded was counter-balanced by the support provided by the bolting together of the consecutive rings of iron.

The size of the cast-iron segments had initially intended to be 18 inches (457mm) in length, but by careful calculation of the stresses that were experienced during the experiment, it was found that the length of the segments could safely be increased to 24 inches (610mm). This decision not only reduced the total cost of the cast-iron lining, but it also reduced the programme of works.

The sheer weight of the segments was 18 cwt (863.0 Kg) meant that erecting them by hand was never going to be an option. As it was imperative that the segments had a tight snug fit and went in the precise position, it meant a mechanical solution to ensure safety and speed would have to be found.

Nuttall's had foreseen this problem and in co-operation with Markham & Ltd of Chesterfield had designed, a machine, operated by compressed air, with a telescopic rotating arm. This was mounted on a lorry with a platform which could be moved from side to side of the lorry. As the lining proceeded, the lorry was moved along the tunnel and after slight alterations had been made, it was possible to standardise the method of putting the tunnel linings into place.

In each segment of the lining two grout holes were provided, each large enough to enable the nozzle of a tube to be inserted, through which liquid cement or grout could be pumped under pressure. Each end of the erected linings had to be sealed, and this was done by the insertion of a 'sausage,' that is, a long fibre bag into which liquid cement was pumped. When this set, the space behind the lining was sealed, with the exception of the vents which were left to permit the escape of air.

A variety of fillings were tested, including cement, injected by means of a cement gun, ordinary grout, injected by compressed air from the grouting pan, and also by the packing by hand of the space behind the lining as it was erected, with pieces of selected rock, and following this, by the injection of grout composed of cement and sand.

As a result of these experiments, it was decided that the last method was the most efficient, as well as the most economical. Sections of the lining in tunnel were grouted under pressure from the bottom upwards, and as the grout completely filled the space immediately behind the segment through which the nozzle was inserted, it rose and began to emerge from a grout hole in one of the segments above. That was evidence that the injection of grout into the space under treatment was adequate. The nozzle was therefore transferred to the section above, and the holes were closed by the insertion of the plugs provided for the purpose.

Chapter 8
Start Tunnelling

The under-river section of the tunnel, which was 5,204 feet in length was constructed between the two shafts and given that this was the longest tunnel of its time, and the geology of the river was not entirely known, there was no certainty that the tunnel would pass through the solid rock. From the data collected in 1886 following the opening of the railway tunnel, it was expected that the tunnel would pass through solid rock. It was known, however, that at one point a glacial channel cut deeply into the rock of the river bed, although its exact width, depth and location could not be determined

In order to avoid an inrush of water from the Mersey into the tunnel workings, it was decided to bore two pilot tunnels. The excavation of the lower of these tunnels was kept about 150ft in advance of the upper pilot tunnel. To give you an idea as to the size, it would not be to dissimilar to the ones currently seen on the Merseyrail Loop system.

The ground through which each pilot tunnel had to be cut was explored by boreholes extending for at least 100 ft in advance of the working face. When the engineers were assured that the conditions ahead were satisfactory, the excavation would move forward into the area previously occupied by the pilot hole. The pilot hole was constructed with drills which operated by compressed air, and enabled the rock to be easily broken up into fragments capable of removal by the waggons brought to the working face.

As work progressed it was decided to use explosives and this halved the time but it was not adopted until a series of experiments had been made to determine the smallest amount of gelignite along with its effect and safety. The charges fired by an electric fuse in order to avoid any undue strain on the surrounding rock. No less than sixty tons of explosives were utilised in the course of the excavation. In all, 147,000 lb. of gelignite were used for construction of the tunnel.

From the roof of the lower tunnel exploratory core bore holes were made upwards in order to discover the character of the overlying strata through which the upper pilot tunnel had to be cut. The system adopted was continued until the two lower headings met in mid-river. The information within the core samples varied according to the direction in which they were driven. The aggregate length of these boreholes amounted to nearly 50,000ft.

In the case of the upward borings they were of the diamond core variety, and enabled an exact record to be made of the contour and character of the rock above the point at which the roof of the full-sized tunnel was to be constructed. As a precaution, the boreholes were carried beyond the actual rock itself and into the gravel & boulder clay of the actual river bed.

When the contour of the rock had been charted and the extent of the glacial channel approximately determined, the number of boreholes in its vicinity was increased. This was to allow the verification of the chart data and to ensure that any necessary precautions could be taken. These diamond core borings were 2 ¾ inch in diameter, and their total length was 4,500 feet. The exact yardage of excavation work which had been accomplished had been recorded daily, and when the workings on each side had progressed to within a few hundred feet of each other, it became possible to state with practical certainty what date the breakthrough would be.

To the layperson, it seems amazing that two gangs of excavators working towards each other underground, through 5,000 feet of rock, can meet at a pre-determined point below ground. The engineer would be amazed if the result were not as planned. Given the accuracy of survey work even at this early stage, the meeting of the tunnels at the pre-determined point and at the pre-determined depth must follow as surely as night succeeds day. That the Liverpool and Birkenhead workings met with such tiny variations from the plan, is merely a proof of the ability and the painstaking care which engineers and contractors alike had devoted to their task.

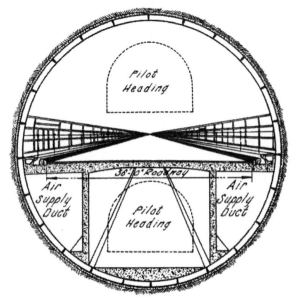

Diagram of Cross Section of Full Size Tunnel
Also showing the relative position of Upper and Lower Pilot tunnels)

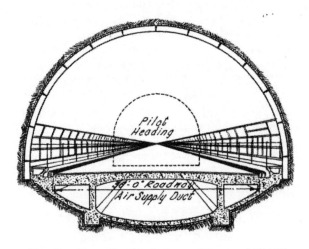

Diagram Showing Cross Section between River and Old Haymarket
Also Kings Square and River Entrance in Birkenhead

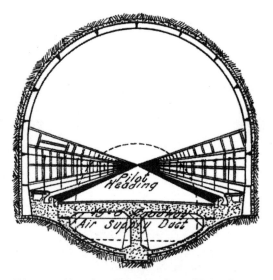

Diagram Showing Cross section of Branches at
New Quay, Liverpool and Rendel Street, Birkenhead

Tunnel Lining

The cast-iron lining of the tunnel is composed of over 100,000 panels and weigh in excess of 80,000 tons. They were manufactured by the Stanton Iron Works Co Ltd. (The site in Ilkeston Derbyshire is now closed with the last cast taking place on 24th May 2007).

    The tunnel segments are bolted together and the panels vary in size. A large proportion of the panels measured 6 ft in length and 2 ft or 2 ft 6 ins. in width, the flanges being 13½ ins. in depth.

At certain points, the tunnel curves and inclines had to have tapered rings formed by special segments. The segments were an exceptional size, it was essential that the greatest care and accuracy should be used in their production. They were required to be cast in special moulds, whilst the utmost care was taken to ensure that the molten metal was of the correct composition and temperature. The time allowed for the moulding was carefully calculated to obviate any imperfection in the finished casting. The panels were then machined to render them interchangeable and also to ensure the tightness of the final fit when they were bolted into their final position.

The importance of this had been stressed by the engineers in that an error of one hundredth of an inch difference in the flange would be fatal in its final quality check. The final check would see a fitting to a master ring to ensure it is within the necessary tolerances before being used in the tunnel.

The nuts and bolts that were required for the tunnel totalled over three-quarter of a million and weighed nearly 3,000 tons. In addition, washers for each bolt added another 130 tons to the overall weight of the materials used in its construction. The bituminous grommets that were used to seal the boltholes totalled over three million, one-third of these being lead-faced. The seal secured the unit to be capable of resisting a water pressure of 200 pounds to the square inch without the slightest danger of leakage.

When the lining of the tunnel had been completed, and the engineers were happy that the linings had been rendered practically impervious to water, the whole of the interior of the tunnel would be covered with a layer of 270,000 tons of the very finest cement. This additional layer would not only smooth the inner surface out but it would also further ensure the water tightness of the lining.

Fortunately, Liverpool was well situated not only to receive goods but geographically close to a good supply of materials. The Welsh Quarries of the Pemnaenmawr & Welsh Granite Company and the Rhosesmor Sand & Gravel Company supplied crushed, screened and washed gravel to McAlpine, both for the tunnel itself and for the ventilation stations which they had constructed in reinforced concrete. Additionally, the gravel was to be used the reinforced concrete road at the King's square, entrance in Birkenhead.

Another source of supply was found in the river itself, for William Cooper & Sons Ltd specialised in dredging the bed of the Mersey. Indeed, its quality was such that it was the only sand used by the contractors for making the concrete they required for the tunnel. The Gresford Sand and Gravel Company Ltd were suppliers of their Gresford Gravel, in a variety of grades.

Within the tunnel, it had been decided that the lower portion of the wall would be protected by a dado. As common, the exact style and materials would not be decided until the late stages. To allow for the materials a sum had been included within the original cost estimates for the works. Whilst the protracted decisions on the dado were underway, the engineers were busy deciding on the final choice of material to form a waterproof layer over the gunite rendering to which the final plaster coating would be laid. The final choice of material to cover the 80,000 sq yards at three gallons per 100 sq feet would be Indasco bitumen Emulsion from the Royal Dutch Shell Group (better known as Shell for their petroleum stations).

Indasco is pure bitumen emulsified with water with the addition of a special stabiliser which prevents flow under heat. The addition of water to the bitumen makes it easy to apply as no heating is necessary, and consequently there is no smell or danger to the workmen. When the material is applied, the water evaporates and the particles of bitumen coalesce, forming a continuous film of bitumen, which is unaffected by water once it has dried out. Moreover, bitumen that has been treated by the Indasco process is in a state that enables plaster and other oily materials to be applied over it and a satisfactory bond obtained. The process involved a single application at 75–80 psi (lb per sq inch) with two De Vilbiss applicators working from a 50-gallon pressure container.

The next big question was what sort of finish is to be applied to the tunnel walls? The committee looked at various finishes in tunnels throughout the world which seemed 'not up to standard' where the Mersey Tunnel was concerned. The Mersey Tunnel was much different from the others as it has the weight of a large river over it and the crushing effects of the water pressure. The engineers had used a cement gun render but this left the surface with a high density but little suction. Throughout the tunnel, there was a humidity of 80 to 100 per cent, which would cause heavy condensation. The super-dense gunite cement rendering they were left with a surface, which was continually running with the water of condensation, thus presenting immense difficulties in the application of any finish. Quite apart from these difficulties and conditions, the finished surface should not be subject to unsightly crazing, so the chosen material had to withstand a high acid condition produced by the fumes from motor and steam vehicles dissolved in the salt and humid atmosphere.

The surface must be easily washed down for such things as grease, soot and other tenacious deposits which would be considerable. Added to these difficulties it was imperative also that the engineers should find a finish to cope with the requirements at as economical a cost as possible, much like todays large projects.

A great many finishes of cement glaze, synthetic and paint types were tested but the chosen material was 'Marplax' marble finish. This was produced by James B. Robertson & Co Limited, of London and it came with a five-year guarantee. 'Marplax' would not craze or deteriorate under fair wear and usage and it incorporated a special plaster base, which was not merely a surface skin. The 'Marplax' finishes to the periphery of the tunnel have been carried out in a pale oatmeal colour, which harmonises with the dado.

One of the dangers of a tunnel is the dreaded influx of water. To alleviate this there is a lot of sealing and re-sealing of the various components that make up the tunnel lining. In the case of the Mersey Tunnel, this was known as the Cementation Process. The process had been patented by Francois Cementation company limited of Doncaster, who's Managing, Director a Mr A. R. Neelands had previously given evidence at the House of Lords on the Mersey Tunnel Bill.

The process consists essentially of boring holes into the porous or fissured strata and injecting a liquid mixture of cement which finds its way into any fissures or cavity under treatment, gradually filling it until the development of pressure on the controlling gauges indicates completion of the injection. Any excess water passes off into the ground or is drawn off by relief holes. The result is a neat cement which sets under pressure and is strong in both tension and compression.

Drilling of the holes within the fissure is done in stages, and in the case of red sandstone, the holes are then injected with chemicals and cement. After the cement, has set the injection cycle is repeated to ensure stability of the structure and no time is lost in the construction process. For the treatment of red sandstone, it is necessary to employ chemicals since the rock is porous, being traversed by fissures of varying sizes down to the finest hair cracks.

It was found during the earlier work in red sandstone that after the larger fissures were filled, the pressures necessary to force the cement into the finer cracks were excessive. In addition, the friable nature of the sandstone was such that frequently the action of the grout caused the formation of dams of fine sand, which prevented further progress of the cement emulsion. To overcome this difficulty the process known today as silicatisation was invented and developed.

This process consists of the preliminary treatment of the ground with chemicals in the case of new red sandstone, silicate of soda and sulphate of alumina. These chemicals combine in the ground to form a gelatinous precipitate. It lines the periphery of the open passages or cavities with a lubricant, enabling cement to pass with little flow resistance and consequently lower pressures, preventing erosion and damming, which in addition permeates the pores of the ground, which can, if necessary, be rendered impermeable.

The tunnellers drilled various holes in the roof to get much needed information on the make-up of the ground below the riverbed. The value of these exploratory drillings was shown when they proved that over a considerable length of the river section a bed of gravel and sand existed between the rock surface and the clay bed immediately underlying the river. As a result of the discovery of this gravel bed, a cementation programme was carried out to consolidate the gravel to enable the tunnel to be driven through.

A sample of the area was taken using a 6-inch diameter core boring was made and showed a perfectly consolidated core. This sort of process is still carried out today to ascertain the depth and make up of say a concrete foundation. As a result of the evidence obtained, however, the tunnel grades were altered and penetration of the gravel bed by the full tunnel section was avoided

The quantities of drilling carried along with the quantities of cement and chemicals used in both the cementation and boreholes works are as follows:

CEMENTATION:
Drilling   430,100 feet
Cement and chemicals injected   3,500 tons
EXPLORATORY BORINGS   57,700 feet

The Stanton Ironworks Company Ltd near Nottingham who manufactured the Queensway Tunnel Segments had many years of experience in this type of work. Its casting operations and the foundry was well equipped for such a large operation. The company were in fact one of the first to supply cast iron tunnel segments and were used in the London Underground Railway. The total number of segments in the tunnel lining is, comprising a total weight of 82,000 tons. Throughout the tunnel there are different sized segments and the majority of which are 6 feet (1.82 metres) long x 2 feet (609 mm) wide or 6 feet (1.82 metres) Long x 2 feet 6 inches (762 mm) wide. The fixing bolts are contained in flanges which are 13.5 inched (345 mm) deep.

The order for the main segments was given to Stanton Company on 8th July 1928 and casting was commenced on 15th July 1928 with output reaching 1,000 tons per week within a month. This was later to increase to 1,250 tons per week as the contract progressed. As the segments were moulded in green sand, and due to their unusually large size a considerable amount of gas was generated during the pouring of each segment. To enable this gas to be expelled 'vented' was completed by hand using moulding boxes. The boxes were handled by overhead electric telpher cranes and pneumatic hoists. During the course of the contract no fewer than 248 types of cost iron pattern were used.

Great care had to be taken to ensure that the finished units were of the correct metal composition along with the casting temperatures and stripping the units from the patterns was correct. And enough time had elapsed from the actual moulding of the unit. Following from this the segments were machined via specially installed milling machines to ensure they were interchangeable. The milling machines were capable of a rate of almost one hundred square inches per minute. This was important as an error of only one hundredth of an inch would have thrown the opposite side of the ring out by around half an inch (12 mm).

The grooves for the caulking were machined out as this was far easier than trying to cast them within the individual segment. The grooves were 1 ¼ inches (318 mm) wide by 1/8th inch (3 mm) deep. Before any segment was dispatched to the tunnel, all segments were checked and this included the flange for absolute accuracy. Once this was complete, the installation of the cast iron segments, some of which were by hand if space was limited. Given the shape and weight of the segments, this was no easy task to complete at the best of times. Once the chambers of the tunnel were wide enough, erectors were positioned to assist in the placing of the cast iron segments. To allow a good work rate and reduce the risk of a collapse, the segments were placed as soon as the space would allow. In difficult ground, the rock was removed with more care and in fewer quantities than before. This was to allow the maximum removal of material but also allow the tunnellers to maintain his safe working area.

The glacial channel of the tunnel was in places 3ft 6 in deep. This does not sound a substantial amount of rock and other material between you and the river, but there were numerous calculations and experience to ensure this was sufficient. The rock in the roof was also supported by the installation of timbers reinforced by steel bars, and the excavation was done in this section by the aid of pneumatic hammers only.

The grouting of the space behind the cast-iron lining was carried out by the method that had been determined in the experimental section and was thoroughly inspected. Any hollowness or suspicions on the lining were thoroughly checked and remedial works undertaken. The grout was inserted at 120 lb psi (Pound per square inch) which enabled it to be injected into the spaces behind the lining segments and any fissures (a long, narrow opening or line of breakage made by cracking or splitting, especially in rock or earth) in the rocks behind.

The boltholes for the segments were made a quarter of an inch larger than the actual bolts. This was to avoid any possibility of water leaking through this tiny intervening space. A dished wrought-iron washer was placed at the head of the bolt, and the nut with a bituminous grummet. When the nut was screwed on the bolt and was tightened, the bolt head flattened out the dished washer, and the bitumen was forced into the space round the bolt in the bolthole.

As an additional sense of security against leakage, joints in the segments varying in depth from one-quarter to one-of caulking in the shape of lead wire. When it was possible the grooves were thoroughly cleaned out and lead wire of the same diameter as the groove was caulked into the bottom of the grooves by pneumatic hammers. The lead wire was manufactured by Perrin Hughes & Co Ltd of Bridge-Water Street Works, in Liverpool. The total amount of lead wire used in the tunnel was in excess of 180 tons, at a length of more than 170 miles.

It is difficult to realise the immensity of this task, but by the aid of a small army of zealous and efficient inspectors, every ring of the lining was thoroughly examined. The necessity for this minute inspection was due to the fact that the next process in the construction of the tunnel was the filling of the space between the flanges of the cast-iron segments with concrete. And it was considered to be imperative that before this was done there should be absolute certainty that water had been excluded from the internal face of the lining.

The engineers at the outset realised that the headings for the pilot tunnels were to begin the point's remote to the Old Haymarket and Chester Street on their respective bearings. The shafts at St Georges Dock and Morpeth dock were to be constructed on a different line to the tunnel. The branch tunnels unlike the main tunnel were to be a semi-circular and lined in cast iron. In the case of the under-river section, the excavation below the road deck was only capable of a shallow invert and was constructed in concrete in the land section of the tunnel. The actual tunnel is 780 feet (238 metres) beyond quay wall of St Georges Dock Parade.

In the areas where the excavation penetrated below the standing water level in the rock, it was necessary to provide a layer of concrete and an efficient damp course of bitumen. This also included the construction of a 12-inch thick inverted concrete arch. The rising gradient of which carried the bottom of the tunnel above the standing water level, waterproofing was no longer necessary. At this stage, it was also possible to reduce the strength of the inverted concrete arch.

At this point, it was realised that the cast-iron linings would not be required for the remainder of the tunnel. This would be a saving in construction costs and was welcomed by the committee. There was a small length of the tunnel that was, therefore, constructed in curved steel joints sections at intervals of 30 inches. The steel has to be protected, which was more important here given the salt atmosphere and rock. The steel in the tunnel was placed above the standing water but within the rock strata, the steel was encased in concrete.

Upon the tunnel emerging from the rock strata, it reverted back to cast iron lining. As the tunnel neared the surface, it had to take into account the stress of the building and structures above. The engineers had to make any necessary adjustments in consideration of the structures safety. Added to this would be all the labyrinth of utilities underneath the surface of the road and pavements

The first and only serious engineering setback to the process occurred on the 29th October 1930. A portion of Dale Street close to the then Police Station collapsed. This was believed to be due to the line of the tunnel crossing of the old fortifications of Cromwellian days. That collapse caused some disruption to street traffic for a considerable time, did not seriously delay operations underground.

As the tunnel proceeded underneath Dale Street in Liverpool, it was initially intended to excavate without closing the street to Traffic. Today this road is an exceptionally busy road through the city. The proposed method was for deep trenches beneath the pavements to the bottom level of the tunnel. The excavations would be supported, as was the roadway along with the utility services by using large timber sections. Upon a detailed examination of the proposals, it was decided not to adopt this method of construction. The decision was made to progress the excavation underground and to be able to achieve this, it would be necessary to design a support for the above ground elements.

The shield would have to not only support rock, which was excavated by the use of explosives and clay. It would also have to cope with any made ground (an area of land that has been made by people, generally through the reclamation or an artificial fill such as landfill) as the tunnel finally rose to the surface.

The final design for the shield was constructed in the tunnel and had an overall diameter of 46 ft 9 ½ in and a length of 12 ft 6 in. To ease movement in such confined, space the shield was mounted on rollers. These were assisted by side headings which were driven in, in advance. In addition, 24 hydraulic rams with a propulsion force of over 2,000 tons helped the forward movement. To protect the men constructing the tunnel, the shield had a canopy projecting a few feet forward of the cutting edge.

Much like today's TBMs, the shield moved forward slowly and as it progressed the cast iron segments were installed into the tunnel. Progression was slow but assured and the tunnellers were protected by the shield. The process was completed up to the junction with both Dale Street and Manchester Street where daylight shone some light on the project. Upon reaching open air the shield was dismantled leaving a cast iron semi-circular layer, with a skin of concrete into the portal arch.

On the Birkenhead Side of the river, there were certain differences in the considerations for the contract and its works. One main difference is that this side did not require a shield to construct the tunnel as both the main and branch tunnels gained ground close to the surface. The Birkenhead side was far cheaper to purchase buildings and demolish any that may be in the way. There were considerations as with Liverpool for structures above ground and the dangers this may have on the above and below ground areas. A trench was cut to a required depth to allow the tunnel to be built and the area above the tunnel backfilled.

The construction of short sections of the tunnel beneath some streets meant that the existing street levels had to be maintained. Within the tunnel the semi-circular arch was replaced by steel girders, spanning concrete retaining walls built in trenches. The steel joists were erected to unite longitudinally the transverse system of girders, and with the aid of concrete and an asphalt damp course, the roof was completed and the surface reinstated. The rock and earth between the retaining walls was not excavated until after the completion of the steelwork.

As the tunnels required branches to the docks, these were constructed in a slightly additional dimension to the main tunnel. This was required for both structural reasons and the traffic management within the tunnel itself. The branches themselves would also need substantial structural support, especially on the Liverpool side given the large buildings directly above.

The first section of the chambers constructed was the reinforced sidewalls which were 5 ft in thickness. A standard method of the construction was the construction of the main rebar (reinforcing bars) that formed the centre core of the wall. Shuttering (temporary walls to hold the concrete and shape the wall) was then constructed to allow for the concrete element of the wall.

At the top of the walls, a proportion of the rebar was left to link in to the next stage of the construction, the roof. Examples of this can be seen in the pictures of the road deck. Once the walls had gone off and the concrete set enough to take the roofing element. This was constructed using steel joists and concrete.

Following the completion of the reinforced steel construction, the removal of the debris is a simple matter. In the lower pilot tunnel the contractor's waggon-railway system was being operated and driven by electric locomotives. Beneath each chamber and along the line of the tunnel the work progressed nicely and chutes were formed in top half of the tunnel into the lower pilot heading. The debris was shovelled into the waggons and removed to either the Liverpool or Birkenhead shaft.

When the excavation reached the bottom of the lower pilot tunnel, the waggon lines were no longer used for the removal of debris, the delivery of the cast-iron segments and other material required for lining the tunnel. As transportation of materials and removal of debris was vital, a temporary roadway suspended from the roof of the tunnel. On this roadway, the waggon lines were laid, and hoists installed at intervals for the handling both of the excavated rock and of the cast-iron segments and other material used in the erection of the lining and of the permanent roadway which divided the upper and lower sections of the full-size tunnel.

At each junction chamber, automatic electric traffic signals were installed to ensure the safety of vehicles emerging from the branch tunnels. The tunnel rises at 1:300 for 167 feet (509 metres) from Birkenhead Quay wall to a gradient of 1:30.

Statistics for Birkenhead (Queensway) Tunnel

Length of Tunnel	2.13 Miles
Total Length of road surface	2.87 miles
Total area of road surface	11 Acres
Width of main tunnel (between kerbs)	36 feet (10.97 metres)
Width of branch tunnel between kerbs	19 feet (5.8 metres)
Internal diameter of main tunnel	44 feet (13.4 metres)
External Diameter of tunnel	46 feet 3 inches (14.1 metres)
Cross section area of main tunnel	1680 feet (512 metres)
Lowest part of tunnel at High Water	170 feet (51.8 metres)
Average cover, rock, and clay underwater	33 feet (10.0 metres)
Weight of rock, gravel and clay excavated	1.2 Million tons
Weight of excavated soil	560,000 lbs
Maximum amount of water from workings	4300 gallons a minute

Both sides of the river had adequate locations to store and dispose of the materials removed from the tunnel and its chambers. Unfortunately, these were not in the immediate vicinity of the tunnel, but were within a reasonable travel distance.

It will also be obvious that prior to the sinking of the shafts, it was necessary for the contractors to make arrangements for the removal of the rock which had to be excavated, as the boring of the shafts and the tunnels progressed. The magnitude of this task will he appreciated when it is stated that the total amount of rock and earth removed during the course of excavation amounted to no less than 1,200,000 tons. The rate of excavation was more than half a ton of rock for every minute between June 1926 and August 1931.

As the tunnel workings progressed, waggon lines were laid to carry the debris from the working faces to the shafts. On arrival at the foot of the shafts, the waggons were hoisted to the surface and their contents tipped into a chute for removal. Heavy haulage is a specialised industry on Merseyside, and a number of haulage contractors existed possessing fleets of large and powerful waggons capable of dealing with the output of debris on each side of the river. Even today, there are a number of Haulage companies, even specialising in cold storage can be found all over the enlarged Merseyside area on both sides of the river and dock areas.

As the excavators worked in shifts through the twenty-four hours of each day, except Sundays, it was necessary for the haulage operations to be maintained during the same period, so that from 10 p.m. on Sunday night until the afternoon of the following Saturday, the work of removal proceeded without cessation. No sooner had one waggon been loaded and driven off than another took its place at the loading base. It must be understood that whilst the contracts for haulage were placed by the contractors engaged, the disposal of the debris was a responsibility undertaken by the haulage contractors themselves.

During the time of the tunnel construction, there was a large land reclamation programme at Dingle and the Liverpool Corporation were constructing a new river wall and embankment at Otterspool. The site was not only fortunate in their planning at the time of the tunnel, but were also capable of taking all the waste.

The main company selected to transport the waste material from the tunnel was Wellington Haulage who used 30 vehicles to remove over half a ton of rocks and other material. The fleet of waggons were accompanied by a motorcycle to assist in the operations. This motorcycle would travel between the tunnel and dump site and check the route was clear and free from obstructions. This would allow the contractor to maximise the time and mileage to full effect.

During night-time, it was only Dingle that would be used as the conditions at Otters pool were deemed too dangerous. However, the Dingle site was to be used but with great care and in addition to this the area would be illuminated with acetylene flares. This is in effect a large drum onto which there is attached a pole and burner which would give out a large flame to illuminate the immediate area.

On the Wirral (Cheshire) side, the disposal of the tunnel waste was less of a problem. Lever Brothers Limited of Port Sunlight Village owned the Storeton Estate, and within this was a large quarry approximately 500 yards in length, and depths of between 25 and 80 foot so was more than capable of taking most of the tunnel waste. The contractor chosen for the removal and disposal of the tunnel material was G & W Dodd who entered into a contract with Lever Brothers to utilise the disused quarry for the disposal of the tunnel waste. One of the reasons this quarry was used, apart from its vast size, was its proximity to the local road network and access to the tunnel.

To move the 453,000 tons of material, Dodd's used 25 Lorries and four steam waggons along the three-mile route night and day to complete the task. Loading was completed at night with the use of electric floodlights and tipping by the use of acetylene flares similar to the Liverpool side of operations. The quarry was also used by McAlpine and Sons who used a fleet of 6 waggons over a period, with the Birkenhead Corporation also using tunnel spoil for what is not called Crush and run as the base for new roads. This Crush and Run is the grey stones you would normally see as the base course for the new road and you will have seen this on a number of occasions but possibly not paid any attention to what it was.

The site of the quarry is now a large expanse of forest and recreational areas known locally as Storeton Woods so even today, the tunnels are being used in more ways that were originally envisaged. The site of 1.3 Hectare (approximate area) was purchased in July 1993 for £80,000 by Storeton Woods Limited, Chester for a full restoration scheme with Merseyside Task Force and a derelict land grant.

As with any construction process, there is a lot of water used in the manufacture of the various materials not to mention the problems with natural water tables. However, a tunnel has all these and the additional problem of tons and at times millions of tons of water pushing down on the project from above. Water as a natural source will always find the course of least resistance and as we live on a planet full of gravity, this will always be downwards.

This ingress of water flowed into the workings from the surrounding rock and general area, no matter how small would have to be dealt with. The sections of the tunnel nearest to the banks of the river, was safeguarded by the cementation and lack of water tables, the river section could not prevent a considerable volume from passing into the pilot tunnels, or a catastrophic collapse.

The shaft at the lowest point of the tunnel gradient on the Liverpool side of the river was 7 ft in diameter and carried the inflowing water to the Liverpool shaft. This shaft had a battery of powerful pumps, which had been installed for the purpose of discharging it into the Mersey. This was to be proved to be a sound policy as the drainage heading was responsible for the influx of two-thirds of the total amount of water which had to be pumped from the Liverpool shaft.

On the Birkenhead side of the river, the shaft dealt with only a small amount of pumping as the surrounding strata was so small that it was decided not to construct a drainage heading on that side of the river. It was decided that the small amount of water, accumulating in the pilot tunnels on the Birkenhead side should be dealt with by pumping. Subsequently, when the river tunnels were joined, gravity took charge and all the water flowed to the Liverpool shaft.

The walls of the main under river tunnel carry the roadway 18 inches below the horizontal diameter of the tunnel. They are 12 ins. thick, and 21 feet between walls giving a clear uninterrupted span. The central space below the roadway and between the shafts was allocated to buses and trams initially, but this was later classed as future vehicle expansion. The curved passages between the walls and the sides of the tunnel were to form the vast ventilation requirements for the tunnel and utilised as air-ducts.

In the sections of the tunnel directly below the roadway, level consists of a shallow invert, and this naturally reduced the height of the required supports. Instead of a continuous wall, a series of columns were erected at intervals of 7 feet. This allowed the space to be fully utilised as part of the fresh air dusts.

The walls were based on concrete foundations which were laid (connected) with the cast-iron segments in the bottom half of the tunnel. When the concrete wall had been carried to the required height, the steelwork for the reinforcement of the concrete roadway, timbers shuttering was placed beneath the steelwork. This shuttering would form the moult into which the concrete was poured from above. This would allow the shape and details of the road to be accurately cast and the concrete held in the exact required position until it had fully set. During the concrete pour, the concrete would have been vibrated to remove any air in the pour and thus ensure there were no weak points (air pockets) in the finished roadway. Upon completion of the setting process (curing), the shuttering would be gently removed.

To give you an example of how complicated cement and it numerous processes can be, I have included a series of FAQs from the Cement Organisation

What is the Difference between Cement and Concrete?

Although, the terms cement and concrete often are used interchangeably, cement is actually an ingredient of concrete. Concrete is basically a mixture of aggregates and paste. The aggregates are sand and gravel or crushed stone; the paste is water and Portland cement. Concrete gets stronger as it gets older.

Portland cement is not a brand name, but the generic term for the type of cement used in virtually all concrete, just as stainless is a type of steel and sterling a type of silver. Cement comprises from 10 to 15 percent of the concrete mix, by volume. Through a process called hydration, the cement and water harden and bind the aggregates into a rocklike mass. This hardening process continues for years meaning that concrete gets stronger as it gets older.

So, there is no such thing as a cement footpath, or a cement mixer; the proper terms are concrete footpath and concrete mixer.

How is Portland Cement Made?

Materials that contain appropriate amounts of calcium compounds, silica, alumina and iron oxide are crushed and screened and placed in a rotating cement kiln. Ingredients used in this process are typically materials such as limestone, marl, shale, iron ore, clay, and fly ash.

The kiln resembles a large horizontal pipe with a diameter of 10 to 15 feet (3 to 4.1 metres) and a length of 300 feet (90 metres) or more. One end is raised slightly. The raw mix is placed in the high end and as the kiln rotates the materials move slowly toward the lower end. Flame jets are at the lower end and all the materials in the kiln are heated to high temperatures that range between 2700 and 3000 Fahrenheit (1480 and 1650 Celsius). This high heat drives off, or calcines, the chemically combined water and carbon dioxide from the raw materials and forms new compounds (tricalcium silicate, dicalcium silicate, tricalcium aluminate and tetracalcium aluminoferrite).

For each ton of material that goes into the feed end of the kiln, two thirds of a ton then comes out the discharge end, called clinker. This clinker is in the form of marble sized pellets. The clinker is very finely ground to produce Portland cement. A small amount of gypsum is added during the grinding process to control the cement's set or rate of hardening.

What Does it Mean to 'Cure' Concrete?

Curing is one of the most important steps in concrete construction, because proper curing greatly increases concrete strength and durability. Concrete hardens as a result of hydration: the chemical reaction between cement and water.

However, hydration occurs only if water is available and if the concrete's temperature stays within a suitable range. During the curing period – from five to seven days after placement for conventional concrete, the concrete surface needs to be kept moist to permit the hydration process. New concrete can be wet with hoses, sprinklers or covered with wet burlap, or can be coated with commercially available curing compounds, which seal in moisture.

Can It Be Too Hot or Too Cold to Place New Concrete?

Temperature extremes make it difficult to properly cure concrete. On hot days, too much water is lost by evaporation from newly placed concrete. If the temperature drops too close to freezing, hydration slows to nearly a standstill. Under these conditions, concrete ceases to gain strength and other desirable properties.

In general, the temperature of new concrete should not be allowed to fall below 50 Fahrenheit (10 Celsius) during the curing period.

What Does 28-Day Strength Mean?

Concrete hardens and gains strength as it hydrates. The hydration process continues over a long period of time. It happens rapidly at first and slows down as time goes by. To measure the ultimate strength of concrete would require a wait of several years. This would be impractical, so a time period of 28 days was selected by specification writing authorities as the age that all concrete should be tested. At this age, a substantial percentage of the hydration has taken place.

Different Types of Cement

Though all Portland cement is basically the same, eight types of cement are manufactured to meet different physical and chemical requirements for specific applications:

Type I is a general-purpose Portland cement suitable for most uses.
Type II is used for structures in water or soil containing moderate amounts of sulphate, or when heat build-up is a concern.
Type III cement provides high strength at an early state, usually in a week or less.
Type IV moderates heat generated by hydration that is used for massive concrete structures such as dams.
Type V cement resists chemical attack by soil and water high in sulphates.

Types IA, IIA and IIIA are cements used to make air-entrained concrete. They have the same properties as Types I, II and III, except that they have small quantities of air-entrained materials combined with them.

White Portland cement is made from raw materials containing little or no iron or manganese, the substances that give conventional cement its grey colour.

Following upon the construction of the roadway, was the construction of the footpaths and service ducts. The footpaths are not intended for the use of the general public, but merely for the safety of the staff in the supervision of the safeguarding of tunnel traffic.

The footpath is protected by a cast-iron kerb through which fresh air is blown for the ventilation of the tunnel. At the points where the tunnels emerge into the open at Liverpool and Birkenhead the roadway is constructed on normal lines, but before this could be done, it was necessary to erect retaining walls up to the point where the tunnel road meets the public road.

To enable these walls to be built it was necessary to dig trenches, in some cases 30 feet in depth, in order to obtain satisfactory foundations. The only place where any difficulty was experienced was at the Old Haymarket, where, in the underlying ground, a great depth of river-mud was encountered, doubtless a relic of the days when the Old Pool reached to this spot. This difficulty was overcome by driving sheet piling along either side of the walls to an adequate depth.

The decision as to the actual surfacing to be installed was the subject of considerable deliberation, as is the case with any groundbreaking and engineering process. It was essential that the finished road surface should be non-skid, easily cleansed and repaired. Experiments were undertaken with various methods and materials available at the time and it was eventually concluded that cast iron should be utilised

Stanton Iron Works Co Ltd were chosen as the preferred contractor for the supply and fitting of the new roadways finish. The area to be covered was in the region of half a million sections of the new proposed cast iron surface. The surface was to be finished with ribbed tinder slices and a studded upper surface. The cast iron segments were attached to the new concrete road by bitumen, which was also supplied by Stanton's. The bitumen was poured onto the concrete in small sections and the cast Iron segments placed on top. Once the bitumen had cooled from its approximate temperature of double boiling point (Fahrenheit) the cast iron sections were effectively stuck to the concrete road.

At one point in the tunnel where it passes over the Mersey Railway at Birkenhead, rubber-paving blocks had been substituted for cast iron to lessen vibration. There are 2,000 square yards of 'Gaisman' patent rubber paving blocks laid onto the concrete roadway. A specific feature of the tunnel roadway was the 'Gaisman' rubber blocks guide drivers passing through the tunnel. As the use of motor cars or even vehicles for that matter was still relatively new to most people and a vast majority of the public had been still using horse and carts. The prospect of people driving through a tunnel two miles long and having a head-on collision would be horrifying for this new revolutionary form of transport.

The Driving test was not actually introduced until 1935, as outlined below.

1930 – Age restrictions and a form of driving tests brought in for disabled drivers. Full licences for disabled drivers valid for a year.

The Road Traffic Act 1930 introduces licencing system for PSVs.

1931 – PSV drivers could be required to take a test, at discretion of Traffic Commissioners.

First edition of the Highway Code introduced.

16 Feb 1934 – Licences for lorry drivers are introduced under the Road Traffic Act, 1934. The licencing authority may require the applicant to submit to a practical test of their ability. The test was initially voluntary to avoid a rush of candidates.

1 June 1935 all persons who had started to drive on or after 1 April 1934 needed to have passed the test. The test cost 37½ pence and the pass rate was 63%. The first person to pass was called Mr Been. There weren't any test centres and examiners would meet candidates at a pre-arranged spot, like a park or railway station.

The tunnel roadway is first divided into two halves, one half for traffic passing from Liverpool to Birkenhead, and the other by traffic passing from Birkenhead to Liverpool. Each half of the roadway is again divided in two by the rubber blocks to separate the lane in which the slow-moving vehicles will travel from that to be used by fast traffic. So far as the blocks in the central line of the roadway are concerned, they were substantial and raised above the setts that would cause inconvenience to any driver who may inadvertently get over the line. This can be seen today on the modern-day rumble strips used on motorways and dual carriageways.

The blocks laid to divide the lines of traffic in each half of the tunnel are level with the cast-iron surface, and serve merely as a visual guide to the driver. As the branch tunnels only accommodate two lines of traffic, they only need a guiding centre line which, similar to the main tunnel, is composed of plain-surfaced amber rubber blocks, at intervals of one foot.

As a modern-day alternative, we know this system as Cats Eyes which were invented by Percy Shaw of Boothtown, Halifax, West Yorkshire. When the tramlines were removed in the nearby suburb of Ambler Thorn, he realised that he had been using the polished strips of steel to navigate when it was reflected by moonlight or street lighting. The name 'cat's eye' comes from Shaw's inspiration for the device the eye shine reflecting from the eyes of a cat. In 1934, he patented his invention (patent no. 436,290 and 457,536), and on 15 March 1935, founded Reflecting Road Studs Limited in Halifax to manufacture the items. The reflective lens had been invented six years earlier for use in advertising signs by Richard Hollins Murray, an accountant from Herefordshire and, as Shaw acknowledged, they had contributed to his idea.

The original Toll Booths on the then four entrances were protected from the vehicles by rubber-coveted concrete kerb units. For the surfacing of the outer gradients leading from the public motorways to the mouth of the tunnel, granite setts were used. The chosen type was Bonawe granite, which is quarried and trimmed on the shores of Loch Etive, in Argyle and Bute on the north-west coast of Scotland, by J. & A. Gardner and Co Ltd of Glasgow. This granite is not only of the most durable quality when subjected to heavy wear, but in addition its surface is such that it is practically impossible to skid on it. Thousand tons of setts were required for the works.

Decompression sickness may not at first be something you attribute to the construction of a tunnel, let alone a sub aqua tunnel. The truth is that in a confined space such as a deep tunnel, decompression sickness is a real problem. Tunnels can be exceptionally long and be very deep and the body will undergo tremendous stress and strain from its day to day requirements on the surface. This is in addition to the numerous occupational hazards that tunnellers have to contend with on a daily basis, tunnel collapse, flooding, fatally injured by machinery in a tight space.

Below is a list of some of the world's longest tunnels and if you think about the length the tunnels will have to go to reach the mid-point (lowest point) before rising again to the surface, this tunnel depth will be considerable.

Gothard Base Tunnel (Swiss Alps) 34.5 Miles long
Thirlmere Aquaduct (UK) 96 Miles Long
Delaware Aquaduct (USA) 85.1 Miles Long
Tunel Emisor Oriente (Mexico City) 38.8 Miles Long

In the original days of tunnelling, the tunnels will be at a great depth and decompression sickness will be a distinct problem. However, todays tunnellers use pressurised machinery and automatic arms to place the tunnel sections so this is not such a problem today. The History of decompression sickness can be traced back to Robert Boyle in 1670 when he demonstrated that a reduction in ambient pressure could lead to bubble formation in living tissue. Then in 1769, Giovanni Morgagni described the post mortem findings of air in cerebral circulation and surmised that this was the cause of death.

In 1840, the recovery of HMS Royal George, had a member of the recovery team (Charles Pasley), to comment that, of those having made frequent dives, 'not a man escaped the repeated attacks of rheumatism and cold'. First documented case of decompression sickness, reported by a mining engineer was observed in 1841. He was described as having pain and muscle cramps among coal miners working in mineshafts which were air-pressurised to keep water out. Between 1870 and 1910, there were at least 25 paralysed caisson workers, all of whom were described as being cold or exhaustion, reflex spinal cord damage; electricity caused by friction or compression; or organ congestion; and vascular stasis all caused by decompression.

Decompression sickness is caused by a reduction in ambient pressure that results in the formation of bubbles of inert gases within tissues of the body. It may happen when leaving a high-pressure environment, ascending from depth, or ascending to altitude. Joint pain ('the bends') accounts for about 60% to 70% of all altitude decompression sickness cases, with the shoulder being the most common site. Neurological symptoms are present in 10% to 15% of decompression sickness cases with headache and visual disturbances the most common symptom. Skin manifestations are present in about 10% to 15% of cases. Pulmonary decompression sickness ('the chokes') is very rare in divers and has been observed much less frequently in aviators since the introduction of oxygen pre-breathing protocols

The term 'caisson disease' (kay-son) means a watertight container that divers or construction workers use under water was first used in 1873 to describe 110 cases of decompression sickness during construction of the Brooklyn Bridge. When a worker comes out of a pressurised caisson or out of a mine, he will experience a significant reduction in pressure. Caissons under pressure were used to keep water from flooding large engineering excavations below the water table, such as bridge supports and tunnels. Workers spending time in high ambient pressure conditions are at risk when they return to the lower pressure outside the caisson if the pressure is not reduced slowly.

The following description of Decompression Sickness is taken from a diver's point of view, as this is the profession most people associate with the medical term. As depth increases, the pressure of the air breathed also must increase. This causes more of the air to dissolve in the bloodstream. The main components of air is a combination of 'oxygen' and 'nitrogen' gases with oxygen being in continuously used by the body, but nitrogen is not used. When a diver ascends, the pressure decreases and the blood can no longer hold all the nitrogen dissolved in it. If a diver ascends slowly, the nitrogen escapes into the lungs and is breathed out harmlessly. But if the diver ascends rapidly, the nitrogen forms bubbles in the blood that can lodge at joints such as the elbow or knee and cause pain. In severe cases, extreme pain causes the sufferer to double over, hence the common name 'the bends.' Symptoms of the bends usually show up within 90 minutes of diving but may take as long as two days. Minor cases cause itching, rash, joint pain, or skin discolouration. Severe cases can see extreme pain at the joints, headache, seizures, hearing problems, nausea and vomiting, back or abdominal pain, vision disturbances, or chest pain.

Minor cases of the bends usually require no treatment; although, a doctor should be consulted. However, severe cases require a hyperbaric (hy-per-bare-ik) chamber, a specialist enclosure that creates pressure to re-dissolve the gas bubbles. The patient is placed under high-pressure conditions, and then the pressure is slowly decreased. Prompt treatment increases the chances for a complete recovery. In cases of a more severe situation, symptoms may be similar to those of stroke or can include difficulty breathing and chest pain. People are treated with oxygen and recompression (high-pressure or hyperbaric oxygen) therapy. Limiting the depth and duration of dives and the speed of ascent can help with prevention.

Late effects of decompression sickness include the destruction of bone tissue (dysbaric osteonecrosis, avascular bone necrosis), especially in the shoulder and hip, which produces persistent pain and severe disability. These injuries occur among people who work in a compressed-air environment and divers who work in underwater habitats. These workers are exposed to high pressure for prolonged periods and may have an undetected case of the bends. Bone and joint injuries may gradually progress over months or years to severe, disabling arthritis. By the time severe joint damage has occurred, the only treatment may be joint replacement. Permanent neurologic problems, such as partial paralysis, usually result from delayed or inadequate treatment of spinal cord symptoms. However, sometimes the damage is too severe to correct, even with appropriate treatment. Repeated treatments with oxygen in a high-pressure chamber seem to help some people recover from spinal cord damage.

Today doctors recognise decompression sickness by the nature of the symptoms and tests such as computed tomography (CT) or magnetic resonance imaging (MRI) sometimes show brain or spinal cord abnormalities but are not reliable, however, X-rays are needed to diagnose dysbaric osteonecrosis.

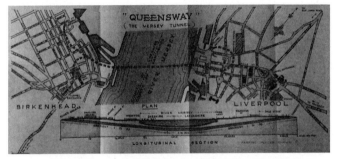

Birkenhead route selection (Wirral Archives)

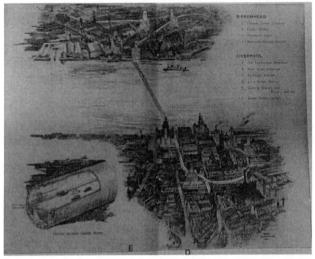

Illustration of the tunnels and its typical section (Wirral Archives)

Birkenhead side
1 Chester Street Entrance
2 Rendel Street Entrance
3 Woodside Ferry
4 Hamilton Square Station
Liverpool Side

A Old Haymarket
B New Quay Entrance
C Exchange Square Station
D Lime Street Station
E Central Station
F James Street Station

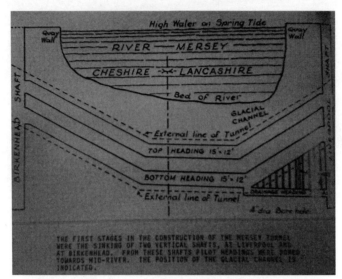

Section through Queensway Tunnel (Wirral Archives)

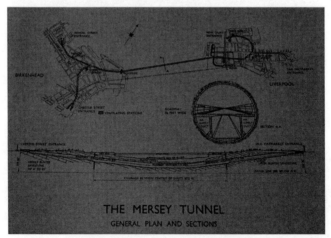

Typical Sections (Wirral Archives)

Starting St George's shaft (Birkenhead Library)

Axe Used by Archibald Salvage (Author)

Tunnel Breakthrough (Birkenhead Library)

St George's Shaft Cementation (Birkenhead Library)

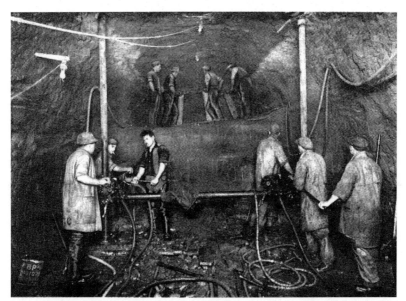
Excavation of Preliminary Tunnel Prior to Pilot Tunnel (Birkenhead Library)

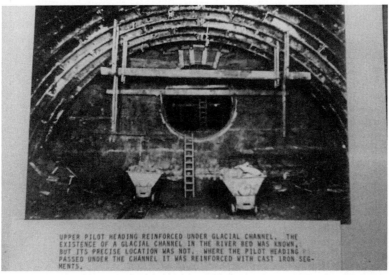
Primary Tunnel (same as London Tube) and Main Tunnel (Birkenhead Library)

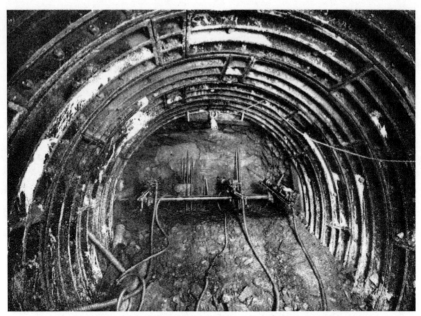
Cast Iron Lining of Preliminary Tunnel (Birkenhead Library)

Rock Face and cast iron Junction
Note Cast iron top left of picture (Birkenhead Library)

Steel and Rib Construction (Liverpool Museum)

General Tunnel Construction (Liverpool Museum)

Tunnel Section Mould (Birkenhead Library)

Tunnel Section Erector (Birkenhead Library)

Erecting Cast Iron Segments (Birkenhead Library)

Cast Iron Lining Over Merseyrail Invert (Liverpool Museum)

Sealing Cast Iron Joints (Liverpool Museum)

Concrete Filling Cast Iron Segments (Liverpool Museum)

Top Section of Tunnel Cast Iron Lining (Birkenhead Library)

Bottom Section Cast Iron Lining (Birkenhead Library)

200 Ton Shield (Old Haymarket) (Birkenhead Library)

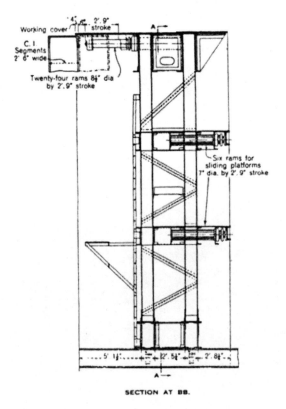

Sections BB through roof shield (Birkenhead Library)

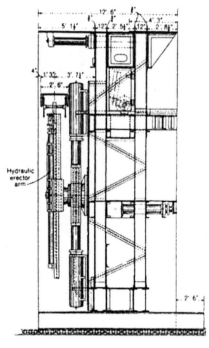

Sections CC through roof shield (Birkenhead Library)

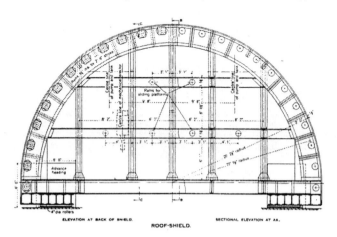

Queensway Roof Shield (Birkenhead Library)

Tunnels and Dock Exits under Construction (Liverpool Museum)

Completed Roadway and Coast Iron Lining (Birkenhead Library)

Birkenhead Entrance Cover Beams (Birkenhead Library)

Section at Rendel Street (Liverpool Museum)

Rendel Street Construction (Liverpool Museum

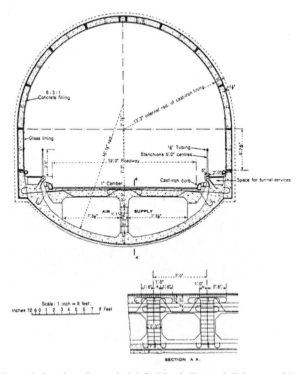
Tunnel Branch Section through 26 ft 6 Inch Tunnel (Liverpool Museum)

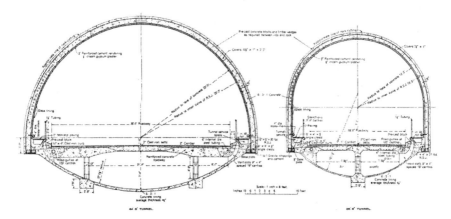

Tunnel Branch Cross Section through Steel and Concrete Tunnel (Liverpool Museum)

Rendel Street Entrance under Construction (Birkenhead Library)

Rendel St Roadway Construction (Birkenhead Library)

Road Deck Shuttering (Liverpool Museum)

Road Deck Steel Awaiting Concrete (Liverpool Museum)

Road Deck Land Section (Liverpool Museum)

Construction of Roadway Side Support Walls (Liverpool Museum)

Road Deck and Supporting Walls (Liverpool Museum)

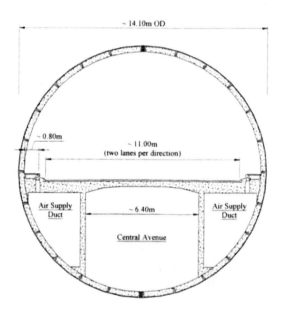

Figure 1: *Tunnel Cross-Section*

Queensway Tunnel Section through Roadway (Wirral Archives)

Queensway Tunnel Cable Laying June 1933 (Liverpool Museum)

Liverpool cut-and-cover arch under construction

Liverpool Cut and Cover for Queensway Tunnel Construction (Wirral Archives)

Side Walkway and Vents (Birkenhead Library)

Side Walkway and Cable Ducting (Birkenhead Library)

Chapter 9
Drainage Services and Ventilation

The fact that the tunnel has been sealed against the incursion of water from the surrounding strata did not suffice to avoid the necessity for the installation of a system of drains. By its very nature, a tunnel will never be waterproof.

In the case of the Queensway Tunnel, there are four large openings (one of which is now disused) which allow the weather to easily enter the complex. This together with the natural gradients to allow traffic to drive from street level to below the river as it has to have a constant downwards gradient at both ends. If there were no drainage solutions, you would see a large body of water collect at the lowest part of the tunnel, which would get bigger over time as there is no sun to evaporate it.

In the event of a fire occurring in the tunnel, the water used by the emergency services must have an outlet, and all this surplus water requires eventually to be discharged out of the tunnel and possibly into the river. The drainage channels which have been constructed are therefore connected to sumps. One of these has been placed at each of the four entrances to the tunnel. There are two others are situated in the vicinity of the George's Dock and Morpeth Dock shafts, whilst the seventh has been constructed at the lowest point of the tunnel beneath the river. The last is the largest, and it is situated at a level below the roadway of the tunnel.

The following table shows the details and character of the pumping equipment at each sump:

Name of Station	No of Pumps	Capacity of each Pump (Gallons per Minute)	Total Head (Feet)	Motor H.P.
Chester Street	2	250	33	7.5
Rendel Street	2	250	33	7.5
Morpeth Dock	2*	300	150	26
Mid River	3	500	220	55
George's Dock	2*	300	150	26
New Quay	2	250	34	7.5
Old Haymarket	2	250	34	7.5
Vent Shaft (Birkenhead)	4**	20	20	
Vent Shaft (Liverpool)	4**	20	20	

NB All pumps have hand and automatic controls but the ejectors in the ventilation shafts are controlled by hand only.

*Provision has been made for a third pump

**These are ejectors, two per shaft

The control of all the pumps is automatic, so when the water in the sump reaches a particular level, the pump will commence the pumping operations. Pumps can be turned off at any time by the control station as the level in the sumps is constantly under observation. If the need arises, any number of pumps can be activated to necessitate the removal of water up to the higher levels and into the river.

Careful computations show that under ordinary conditions one pump in each station will be adequate. The additional pumps are intended to provide that margin of safety which is considered requisite in the event of abnormal conditions arising. It has been calculated that under no circumstances would the full pumping equipment fail to discharge from the tunnel any quantity of water that could possibly accumulate. Even in the event of a breakdown, there are sufficient pumps to clear the tunnels.

Consequential street works involving an expenditure of around £107,000 (£6,671,482.10) were carried out by the department's workmen, from Liverpool's City Engineers at the Old Haymarket approach to the tunnel to reconstruct the Dale Street sewer, which was 6 feet (1.82 metres) high and 4 feet (1.22 metres) wide, and 27 feet (8.23 metres) below the road. A small problem arose when it was discovered that a small 900 feet (274.32 metres), section between Stanley Street and Hatton Garden, was found to be below the site of the arch of the new tunnel. A position under the north footpath of Dale Street was found to be suitable for the new sewer. The Manchester Street sewer originally discharged into the old Dale Street sewer, but when the position of the Dale Street sewer was to be altered to the north side of the roadway all connections from Manchester Street were cut off. And it was necessary to make a complete change in the direction of drainage; the sewer was diverted at the junction of Manchester Street and Victoria Street into the Whitechapel sewer.

As a consequence of the tunnels, an area of approximately two acres around the entrances had to have their pavement levels altered. This required a large street works engineering project to be undertaken in a busy city centre at a cost of around £79,000 (£4,925,673.70). Not the best of projects to be undertaken at any time of the year, and although, the costs seem minimal, to today's comparative costs and machinery to be used, this is a phenomenal amount for the intended works. The work would not only cover the realignment of the pavements, but tram tracks (30 points, 70 crossings), and junctions, of the five main routes at the Old Haymarket but roads and other surfaces.

The work was carried out during office hours and as everyone knows the noise let alone the inconvenience to cars and pedestrians can be infuriating, especially if the project is a prolonged period one. The works were also to see the demolition of the public toilets in the Old Haymarket but construction of new ones behind the boundary wall of St Johns Garden's at a cost of £7,500 (£467,627.25).

The case for both sides of the river to consider in the advance tunnel works was to avoid the dislocation of the public services. This would include public sewers, gas and water mains, electrical, telephone and telegraph cables. It was necessary to involve the relevant companies to re-route the services before the tunnel workings could be carried out. This was no easy task, especially in connection with the sewerage system.

As with Liverpool, Birkenhead would see its fair share of road works to be carried out by the Borough Engineers Department. Similar to Liverpool, Birkenhead entrances were to see pavement and road levels changed to allow the tunnel entrances to be constructed. There was also the consideration of the trams and the nearby train lines and junctions for Cammell Lairds Ship Builders.

At the Rendel Street entrance, the lowering of the sewers was a bit of a problem in that the new level was on or very close to sea level. As the entrance was close to Egerton Dock and this the dock and river system, the new works would have to ensure the inflow of water from the tidal river could be prevented. New sewers were a variety of sizes ranging from 6 feet (1.82 metres) high by 4 feet (1.22 metres) wide and smaller ones of around 2 feet (610mm) in diameter. Today we are used to seeing the circular concrete ones of varying sizes piled up along the road and motorway system.

The ventilating plant and the buildings exceeded £1,000,000 (£62,350,300.00). Dr John S. Haldane, president of the Institute of Mining Engineers, and an acknowledged authority on gases and ventilation stated that he had examined the ventilation proposals, and was satisfied with them.

It would be absolutely impossible, he added, for carbon monoxide to gather in pockets in the tunnel, because it would be swept away by fresh air under control at every point along the tunnel.

Mr Basil Mott, the designer of the tunnel, and Mr J. B. Lister, of the Sturtevant Engineering Company, also spoke confidently of the adequacy of the proposed ventilation.

"If there should be a breakdown," said Mr Mott, "when the tunnel was full of vehicles from end to end, the quantity of air provided for would be sufficient."

"As to cost," he said that, "he had allowed £840,000 (£52,374,252.00) in his estimates for contingencies."

As welcome as the internal combustion engine was, it has its downside in the deadly and odourless gas known as carbon monoxide. The Queensway Tunnel was to be used by thousands of vehicles, which would emit this deadly carbon monoxide, that it became necessary for the Mersey Tunnel Joint Committee to sanction the expenditure on ventilation.

The committee sent Mr B. H. M. Hewett, at that time the Engineer-in-Charge, together with Mr J. E. Lister, of the Sturtevant Engineering Company, who had given evidence before the Parliamentary Committees, on a mission to investigate the actual conditions in the Holland Tunnel in New York and others which were used by motorcars. It was realised that this question was one of supreme and vital importance, and when the Holland Tunnel had been opened it was decided that full advantage should be taken of the experience which had been gained in the working of that tunnel.

The Holland Tunnel was begun in 1920 and completed in 1927. The tunnel is named after Clifford Milburn Holland (1883–1924), Chief Engineer on the project, who died before it was completed. Tunnel designer Ole Singstad finished Holland's work. The tunnel is one of the earliest examples of a mechanically ventilated design using 84 fans, in four ventilation buildings to create a floor to ceiling airflow across the roadway at regular intervals, via systems of ducts beneath and above the roadway. The fans can completely change the air inside the tunnel every 90 seconds. A forced ventilation system is essential because of the poisonous carbon monoxide component of vehicle exhaust, which constituted a far greater percentage of exhaust gasses before catalytic converters became prevalent.

The tunnel consists of a pair of tubes, each providing two lanes in a 20-foot (6.1 m) roadway width and 12.5 feet (3.8 m) of headroom. The north tube is 8,558 feet (2,608 m) from end to end, while the south tube is slightly shorter at 8,371 feet (2,551 m). Both tubes are situated in the bedrock beneath the river, with the lowest point of the roadway approximately 93 feet (28.34 m) below mean high water. The fans have a capacity of 2.5 million cubic feet of fresh air per minute and deal with the 2.5 parts of carbon for every 10,000 parts of air.

It was in the autumn of 1930 that the engineers and the committee were alarmed by a report from America that a number of people had been 'gassed,' though not fatally, by carbon monoxide fumes in an inadequately ventilated land tunnel at Pittsburgh. This was quite shocking news for the committee and the Pittsburgh Tunnel was shorter than the current Queensway Tunnel so a solution had to be found quickly.

The principal road tunnels of America were visited, and all available pertinent data studied. The conclusion of the trip was thoroughly investigated by experts, such as Dr Haldane and Professor Douglas Hay, in conjunction with the engineers. These investigations and arrangements took up much time, and it was not till February 1932, that the additional cost involved was made known to the public. The first estimate of the deficit was £1,500,000 (£93,525,450.00) including nearly £1,000,000 (£62,350,300.00) for ventilation buildings, ducts, and machinery. But, it was discovered later that even that large sum would not cover commitments from which there was no escape, in respect of not only the revised ventilation plans, but other liabilities not provided for in the previous estimates.

Additional capital was found to be required to the amount of £2,046,000, (£124,299.786.86) the chief constituent items being:

Works	£664,000 (£41,400,599.20)
Capitalised interest	£600,000 (£37,410,180.00)
Land and buildings	£305,000 (£19,016,841.50)
Street works outside tunnel entrances	£150,000 (£9,352,545.00)
Contingencies on ventilation, ducts, equipment, lighting, etc.	£75,000 (£4,676,272.50)
Extra cost of foundations of George's Dock station	£54,000 (£3,366,916.20)
Additional for ventilation Buildings	£40,000 (£2,494,012.00)
Cost of Parliamentary Bills	£37,000 (£2,306,961.10)
Share of cost of reinstating river wall	£5,700 (£355,396.71)

The estimate of £5,222,000 (£325,593,266.60) placed before Parliament in the second Tunnel Bill was thus increased to £7,268,890, (£441,603,850.31) and later to £7,723,000 (£481,531,366.90). As the law stood, the 'extras' were a liability exclusively of the ratepayers of Liverpool, as Birkenhead's liability was limited under the provisions of the Mersey Tunnel Acts to the product of a 4d (£1.04) rate (increased by the operation of the Derating Act to £5 1/3 d (£316.43). It became the duty of the Tunnel Committee to discover ways and means of extrication from a most unfortunate financial situation.

The question to be determined by this experiment was which of the three systems of ventilation under consideration would be not only the most effective, but sufficiently effective to guarantee the users of the tunnel against danger or discomfort.

The systems may be briefly described as:

The Upward Transverse System

Fresh air is blown into large air ducts situated beneath the roadway. From these ducts, the air enters the tunnel itself at roadway level, and the orifices from the air-supply channel are so arranged that equal distribution of the fresh air is obtained throughout the entire length of the tunnel. This effect is achieved by graduating the size of the orifices through which the fresh air enters the tunnel. Those situated where the air pressure is greatest being small, whilst the remainder are enlarged according to the reduction of pressure as the air passes along the supply tube. When the Upward Transverse system is installed, it is necessary to construct below the roof of the tunnel a duct large enough to carry away the impure air, and the volume of this must, of course, be equal to the amount of pure air blown in by the fans

The Downward Transverse System

The Downward Transverse, works in exactly time opposite way. The pure air is blown inn through the ceiling of the tunnel and the impure air extracted at roadway level.

The Upward Semi-Transverse System

Pure air is blown into the tunnel through the ducts under the roadway and at roadway level, as described in connection with the Upward Transverse system, whilst the impure air is drawn along time traffic space to exhaust chambers above the roof of time tunnel at each ventilating station, beneath each of which a slotted ceiling has been constructed in the roof of the tunnel. The size of the slots or apertures in the ceilings are permanently fixed, and they avoid the danger of a fierce draught whilst balancing the amount of air extracted with that which enters the tunnel at roadway level

The committee could not decide on any particular system, so they decided to undertake large-scale tests within the Queensway Tunnel itself. A section of the tunnel between the Birkenhead entrance and Hamilton Square, Birkenhead, had been completed to such an extent that would allow the tests to be carried out. This section was isolated from the remainder of the tunnel by the construction of a temporary brick bulkhead at each end, whilst a removable ceiling was fitted with the necessary adjustable openings.

For the purpose of these experiments, immense preparations had to be made. Temporary blowing and exhaust fans were installed, each having a normal capacity of 300,000 cubic feet of air per minute. Then large quantities of petrol were ignited and bales of damp hay and straw were set on fire to produce dense clouds of smoke, whilst smoke candles were burned to enable the observers to discover in detail time behaviour of the air currents. A steam boiler capable of producing heavy clouds of smoke was also moved through the thousand-foot section of the tunnel, whilst in addition, recording instruments were utilised to measure the low of air both in the ducts themselves and in time tunnel.

The experiments were carried out over a period of several weeks and attended by both Dr Haldane and Professor Douglas Hay, as well as by the representatives of the Sturtevant Engineering Co, Ltd and Walker Brothers (Wigan) Ltd.

When the results were finally analysed, the engineers recommended the Mersey Tunnel Joint Committee to adopt the Upward Semi-Transverse system of ventilation. This abolished the necessity for the construction of an exhaust duct, which would have had to be built below the roof of the tunnel, and so considerably reduced the capital cost of the ventilation system.

Once the final decision had been made of the ventilation system, work on its design and installation could begin. This would also allow the mechanical and electrical engineers to complete their designs for the respective installations. Once this was complete or at least basic designs conceptions had been completed, it would allow Mr Rowse to start his Ventilation shaft designs, the sites of which had already been chosen and the land acquired.

Mr Herbert B. Rowse F.R.I.B.A, who had trained at the School of Architecture at the University of Liverpool. Few people will ever realise the magnitude of the problems which he had to solve. There were no precedents to guide him in the simple, masterly and impressive buildings with massive shafts that were to be prominent features in the skylines of Liverpool and Birkenhead.

Owing to the delays in the determination of the actual choice of system, the engineers found it impossible to complete the design for final instructions to the contractors. Obviously, it was an essential preliminary to the preparation of the designs for the buildings which were to form the ventilation stations that a layout of the machinery to be housed in each building should be obtained. Once this had been completed a task which occupied nine months it became possible for the architect to devote his mind to the designing of the buildings.

There are six ventilation shaft building locations which would be up to 210 feet (64.00 metres) high and would need to discharge the vitiated air at least 50ft (15.24 metres) away from the intake are in the following locations:

1. George's Dock, which contains, the plant, main administrative and control room
2. North John Street, and New Quay, on the Liverpool side
3. Woodside, Sidney Street and Taylor Street, at Birkenhead.

Mr Rowse had a few of the basic questions answered such as the size of the fans and type of machinery to be used. He did not have the answers to how to house them, air was to be taken into the buildings to be blown down into the tunnel, the vitiated air was to be discharged without causing any annoyance to neighbouring buildings, the machinery was to be prevented from disturbing the neighbourhood with noise and vibration, and, lastly, what the character of the external design was to be.

All of these areas had to be answered and the buildings designed in a short space of time to allow the construction and completion by the time the tunnel was opened.

One or two problems almost solved themselves. For instance, all the machinery was to be controlled from one centre in the George's Dock building. No operating staff would be needed permanently in the other buildings, and therefore natural lighting would be unnecessary. So windows could be dispensed with, and that, of course, helped to keep noise and vibration inside the buildings. A further point against windows was that there was bound to be taken into the buildings, with what passes for fresh air in our modern cities, large quantities of soot, and this would very quickly have blackened the windows on the inside.

Some of the other more detailed problems were more of a hindrance than a help. Once a problem was solved, it created more problems to be solved. On similar such tunnels such as that on the Hudson River in America. The air is taken in through the roof and not the vertical openings in the walls, which could also prove ugly for the final building design. The side vents would also see the air travelling out of the building at speed and could possibly cause hammering of the adjoining buildings windows.

Thought had to be put into the possible future elements of the adjoining sites and development.

Within the ventilation buildings, there is an inner building of brick, separated from the outer building by an insulating air space. The driving machinery is insulated from the main structure by a layer of absorbent cork insulation. This insulation is extended to all supports for electrical cables within a certain distance of each motor.

Some of the sites were a little restricted in that the general construction and as the fans could be 28 feet (8.53 metres) across and the fan casings (of which there is between four to six in each building) are around 50 feet (15.24 metres) in diameter and 10–16 feet (3.04–4.87 metres) wide. Associated with this is the crane that would be necessary to lift these items into position. The installation of this equipment was commonly called 'tucking in' which, given the tight spaces, would be quite an appropriate description.

The George's Dock station Mr Rowse was given the additional problem of not only working next to the distinctive Pier Heads Three Graces, but all the machinery was to be in duplicate, to guard against any possibility of a failure in ventilation. This led the way to symmetrical buildings with nicely balanced elevations. The only exception to the balanced elevation treatment occurs in the building at Taylor Street, Birkenhead, where a somewhat awkward site demanded a different arrangement.

George's Dock's three main elements are the blowing-fan chamber, the exhaust-fan chamber, and the switchgear room. 'Fresh air is drawn by suction into the eyes of the blowing fans and expelled into the shafts which connect with the fresh-air supply ducts beneath the tunnel pavements. The exhaust outlets in the tunnel are connected to ducts, which end in the exhaust chambers. The switchgear for the electrical driving machinery is accommodated on a mezzanine floor between the fan chambers or an adjacent room.

Requirements of the kind found in these buildings had not been met before, and they produced conditions entirely different from those met with in normal building construction. For instance, there was a tremendous weight of concrete, in fan casings and air ducts, which had to be supported high up in the buildings. The huge fan casings themselves are constructed in reinforced concrete 9 inches (228.6 mm) thick, and the air ducts, which rise to the full height of the buildings, are also of the same material but only 6 inches (152.4 mm) thick. All this meant new problems for the architect. The beams and stanchions had to be specially spaced in order to carry the heavy fans, driving units and switchgear, while loads were brought on to the steel framing by electrical and mechanical machines and beds, in addition to the considerable weight of the concrete casings. The walls and towers were also to be supported.

In the George's Dock building the architect's devices for supporting all these loads necessitated the use of two thousand tons of steelwork; in all six buildings about six thousand tons were used, the whole of which was supplied by Messrs. Redpath, Brown & Co Ltd of Trafford Park, Manchester, under the direction of the architect.

The Woodside building, stands only 19 feet (5.79 metres) from the river wall, is in many ways the most remarkable and striking of all the ventilation buildings. The site is extremely restricted accommodates a building only 40 feet (12.19 metres) deep, with a width of less than 100 feet (30.48 metres) and a height of 210 feet (64.0 metres). So much machinery had to be accommodated in a much-restricted building that a complicated and baffling set of problems had to be resolved by Mr Rowse. This caused him so much heartache that he was on the verge of despair. Besides the great weight of machinery and fans the question of wind pressures had especially to be considered in connection with a building of such a height and so near to the river.

The inside had to be so full of machinery that there was no room for intermediate steelwork. The steelwork was to support the pressures of the equipment imposed upon it. Eventually, after weeks of the most careful and anxious study, the problem was solved by arranging the four corner stanchions on a double principle with cross bracing and running the total height of both the building and the tower, thus making them completely rigid in themselves, and enabling the structure to resist the enormous strains set up by wind pressure. Another difficulty arose from the fact that the ground on the riverbank was so poor that the foundations of the building had to be of massive steelwork construction, carried on concrete-filled shafts, taken down 40 feet (12.19 metres) to rock level.

Once the problems of the insides of the buildings had been solved, it became necessary to consider their outside shape and appearance. The heights of the buildings had been determined by a symmetrical arrangement of the fans and machinery was dictated by the fact that this was all in duplicate, and so as a rule there followed from this an equally symmetrical arrangement of the bulk of the building as seen from outside.

The George's Dock building is one of symmetrically disposed masses placed axially with the dome of the dock board building, which lies between it and the river. The Woodside building is equally well balanced, and this when completed will be a most impressive sight on the edge of the river

From considerations of neighbourliness towards the other buildings in the business quarters in which they stand, the George's Dock and North John Street buildings are faced with Portland stone. The absence of windows offered the architect unusual devotional problems. Under the direction of Mr Rowse, and assisted by a Liverpool sculptor, Mr Edmund C. Thompson, with whom Mr George T. Capstick is associated, they have managed to concentrate on the ornate structures and decorative features of the ventilation buildings.

The brick buildings have strong expressions of approval from architects, and the design has produced some of the finest brickwork in Europe. The bricks were supplied by Ames & Finnis, and the greatest care was taken in their selection, both from the standpoint of size and that amount of variation in colour which was necessary to enable the brickwork to present an interesting and bright appearance. The bricks are used in different ways to provide relief from what might have been the monotony of plain wall surfaces, but there are large areas, of feature brickwork along all elevations.

The reinforced concrete work on the ventilation buildings demanded craftsmanship on the finest scale. Everything had to be perfectly smooth and with no irregularities on its surfaces. The margin of error of the interiors of the fan chambers was set at a maximum of one quarter of an inch. It was also possible to construct in reinforced concrete such intricate structures as the great air ducts and the fan chambers within the limits set by the mechanical requirements, and that the smooth finish necessary to reduce friction was obtained in concrete without the addition of any sort of rendering. To produce the finish, the Trussed Steel Concrete Co Ltd of London, used carefully graded and proportioned aggregates.

One of the problems confronted both the engineers and the architect with the installation of the ventilating machinery was the great weight, and the difficulty of placing the various sections in their allotted positions in such tight spaces. Provision had also to be made for handling these massive loads at any future period when alterations or repairs might have to be needed. The difficulties were met with the aid of Herbert Morris, Ltd who installed an overhead monorail system with travelling lifting gear to allow maintenance of the equipment.

As an example of the tight working conditions, the rotors had to be lifted 85 feet (25.908 metres) and moved in a transverse direction and through a tight opening to allow their final installation onto the main bearings. The Morris lifting equipment and monorail system allowed full access in emergencies such as changing the bearings of the fans. The stand-by cranes are, therefore, designed to be ready at a moment's notice, no matter how many months they may have been standing idle. The sustaining mechanism is automatic and holds the suspended load as long as necessary.

The tunnel would not be opened for traffic until Mr Rowse was able to guarantee that the ventilating stations would be completed and until the engineers could also guarantee that the machinery would not only be installed, but that it would be functioning properly. As the installation of the machinery could not be started until the buildings were sufficiently completed.

George's Dock and Woodside presented a serious problem in that they had to contain the intricate equipment required for the safeguarding and control of traffic. It was extremely gratifying to all those responsible that the various contractors completed their work to scheduled time, and in several notable cases, such as in the construction of the blowing fans for ventilation in the George's Dock station, the work was done in quite considerable periods ahead of the agreed programme. At the close of this great enterprise the architect was able to report a very high standard of excellence in the quality of the work of the various trades concerned.

Steel Suppliers for All six Ventilation Buildings

Redpath, Brown & Co Ltd of Manchester. But few amongst those who have seen the frameworks of steel girders rise into the sky, a task which had been undertaken before the fabricated steel had been transported to the sites.

When the plans had been finally passed, it became necessary to construct models of every piece of steel that was to form part of the structure. Then the steel had to be fabricated to the required dimensions. Before it left the works, each section was marked to correspond with the working drawings supplied to the erecting staff. The speed with which the framework of these buildings was erected represents a triumph of organising ability comparable only with the efficiency with which each building was completed.

Construction of the Ventilation Buildings

- W. Moss & Sons Ltd built the North John Street station
- Henry Boot & Sons Ltd of Sheffield, the New Quay and Taylor Street stations.
- John Mowlem & Co Ltd of London, the George's Dock station
- Sir Robert McAlpine & Sons, of London, the Sidney Street and Woodside stations.

Electrical Installation

- George's Dock was undertaken by Higgins & Griffiths Ltd London.
- North John Street and Taylor Street by John Hunter & Co Ltd of Liverpool.
- New Quay and Sidney Street by Electric Power Installers Ltd of Liverpool.

Plumbing Works

R. W. Houghton Ltd of Liverpool, were entrusted with contracts for the plumbing work required at George's Dock, North John Street, New Quay, Sidney Street and Taylor Street ventilating stations.

Door Suppliers

The Birmingham Guild Ltd supplied special double doors to the fan chambers in all the ventilating stations. These were designed and fitted as an additional precaution against the transmission of sound and vibration. The outer doors were of hollow steel framing, whilst the inner doors took the form of hollow wood shutters, all being firmly clamped against vibration.

Finishes and Final Coatings

James B. Robertson & Co Ltd of London, whose 'Marplax' finishes were selected for the final coating of the tunnel interior, were equally successful in obtaining a contract for the supply of the final finishing surface of the walls of the fan chambers and switch-gear rooms, in which all exposed concrete surfaces have been treated with 'Stipplecrete', a glazed, stippled, impervious finish.

Quantity Surveyors for All six Ventilation Buildings

Mr W. H. Law, of W. M. Law & Son, 26, Exchange Street East, Liverpool, and the Chartered Surveyors' Institution and Member of the Quantity Surveyors' Committee of the Chartered Surveyors' Institutions, acted as quantity surveyor for the six ventilation buildings for the tunnel.

He was responsible for measuring up Mr Rowse's drawings of the buildings for the quantities of material and labour required for their erection, and his figures were issued to the various firms tendering for the work to enable them to submit their competitive estimates of cost

To enable the ventilating plant to deal efficiently with the widely varying traffic density and atmospheric conditions, that may be anticipated within the tunnel at different times of the. The sixteen fans from Walker Brothers are from their Indestructible range. The impellers are essentially cast-iron which carry two mild-steel discs between them. Between these discs are secured mild-steel plates which project outwards and form the arms supporting the fan blades?

The bosses and discs are held together by turned bolts and nuts and secured to the shaft by keys, while the steel blades are attached to the arms by steel angles. The shafts are of mild steel, the largest being 18 inches (457 mm) in diameter. Each shaft is carried in two pedestals, with ring-lubricated sleeve-bearings, which are mounted on girders placed in the fan inlets. The propellers are 21 ft–25 ft, (6.4–7.6 metres) and 28 ft (8.53 metres) in diameter, and will run at various speeds up to 62 rpm.

The fourteen Sturtevant type VC fans were specially constructed to ensure quiet operation. The impellers are built up on circular steel discs, and carried by two cast-steel half-hubs, mounted on forged-steel shafts. The blades, which have a considerable backward curvature, were pressed in dies to give a uniform shape and thickness, and are riveted both to the centre plates and to the outer shrouds, the latter being coned to give as nearly a streamline low as possible to the air passing through the impellers

All the fan casings consist mainly of reinforced concrete, but in the case of the Sturtevant fans, the casings are provided on each side with inlet panels of ¼ -inch (6 mm) steel plate and angles. These panels are carried on frames of 7in. x 3 ½ in (175 x 28.5mm) steel channel built into the casings. They are made removable, so that the impeller can be withdrawn, and support the air-inlet cones, which have been shaped as the results of experiments to give optimum efficiency under normal working conditions.

The thirty Metropolitan-Vickers type RS squirrel-cage pattern ventilating motors are supplied with energy at 400 volts. The motors are of very rugged construction and have large diameter shafts. They are designed to run quiet and minimum maintenance. All the motors driving exhauster fans are separately pipe-ventilated through a duct from the fresh-air shaft.

The hydraulic couplings between the motors and the reduction gears are similar to the original device from which the well-known fluid-flywheel transmission gear was evolved. The advantage, when the output of the fan is regulated, is that considerably less electrical energy is taken by the motor. This also enables the motors and gears to pick up the load smoothly and without shock, a feature of importance when it is realised that the fans vary in diameter from 7 ft 6 in. (2.31 metres) to 28 ft, (8.53 m) and in weight from 3.35 to 23.75 tons. The couplings give an advantage when they make it possible for the motors to be started without load, and thus make possible the use of electrical starting gear, which reduces the starting current taken from the supply system.

To eliminate vibration and noise, extensive use was made of cork mats and packing supplied by Absorbit Limited of London. The concrete foundations carrying the motors and gears rest on 'Korfund' mats, two inches (50.8 mm) thick, and the sides of the foundations are insulated by 'Korsil' packing, two inches (50.8 mm) thick.

Similar treatment has been given to the foundations of the contactor cubicles and transformers and also to the supports for the cables for a distance of 30 ft (9.14 metres) from the main driving motors.

Liverpool Ventilation Station	Exhaust Units	Blower Units
Georges Dock	1 No 400HP and 735 rpm 128HP and 485 rpm 1 No 400Hp and 735 rpm 59.5HP and 365rpm Both motors drive a 28ft (8.5m) diameter fan, with a capacity of 599,000 cubic feet of air per minute through 38 inch (965mm) nominal diameter hydraulic couplings and double reduction gear	1 No 350HP and 735 rpm 113HP and 485 rpm 1 No 350HP and 735 rpm 52.5HP and 365 rpm Both motors drive a 25 ft (7.6m) diameter fan with a capacity of 496,000 cubic feet of air per minute through 35 inch (889mm) nominal hydraulic coupling and double reduction gear
New Quay	1 No 40HP and 730 rpm 12.5HP and 485 rpm 1 No 40Hp and 720 rpm 6HP and 365rpm Both motors drive a 7ft 6inch diameter fan, (2.28m) with a capacity of 92,000 cubic feet of air per minute through 23	1 No 55HP and 735 rpm 17HP and 485 rpm 1 No 55HP and 725 rpm 8HP and 365 rpm Both motors drive a 9 ft (2.74m) diameter fan with a capacity of 4145,000 cubic feet of air per minute through

		inch (584 mm) nominal diameter hydraulic couplings and single reduction gear	26 inch (660 mm) nominal hydraulic coupling and single reduction gear
North Street	John	1 No 230HP and 735 rpm 75HP and 485 rpm 1 No 230Hp and 720 rpm 35HP and 365rpm Both motors drive a 28ft (8.5m) diameter fan, with a capacity of 522,000 cubic feet of air per minute through 32 inch (812mm) nominal diameter hydraulic couplings and double reduction gear	1 No 115HP and 730 rpm 38HP and 485 rpm 1 No 115HP and 730 rpm 18.5HP and 365 rpm 1 No 65HP and 730 rpm 23HP and 485 rpm 1 No 65HP and 725 rpm 11.5HP and 365 rpm The two 115HP motors drive a 23ft (7.0m) diameter fan with a capacity of 312,000 cubic feet of air per minute through a 29 inch (736mm) nominal diameter hydraulic coupling. The two 65HP motors drive a 23ft (7.0) diameter fan with a capacity of 234,000 cubic feet of air per minute through a 26 inch (660mm) nominal diameter hydraulic coupling and through double reduction gear.

Wirral Ventilation Stations	Exhaust Units	Blower Units
Woodside	1 No 430HP and 735 rpm 139HP and 485 rpm 1 No 430Hp and 735 rpm 63.5HP and 365rpm Both motors drive a 28ft (8.5m) diameter fan, with a capacity of 641,000 cubic feet of air per minute through 38 inch (965mm) nominal diameter hydraulic couplings and double reduction gear	2 No 175HP and 735 rpm 57.5HP and 485 rpm 2 No 175HP and 735 rpm 27.5HP and 365 rpm The above motors drive a 21 ft (6.4m) diameter fan with a capacity of 280,000 cubic feet of air per minute through 32 inch (812mm) nominal hydraulic coupling and double reduction gear
Sidney Street	1 No 310HP and 735 rpm 95HP and 485 rpm 1 No 310Hp and 735 rpm 41HP and 365rpm Both motors drive a 14ft (4.26m) diameter fan, with a capacity of 577,000 cubic feet of air per minute through 35 inch (889mm) nominal diameter hydraulic couplings and double reduction gear	1 No 120HP and 730 rpm 36HP and 485 rpm 1 No 120HP and 730 rpm 16HP and 365 rpm 1 No 35HP and 730 rpm 11HP and 485 rpm 1 No 35HP and 725 rpm 5.5HP and 365 rpm The two 120HP motors drive a 13ft 6-inch (4.11m) diameter fan with a capacity of 350,000 cubic feet of air per minute through a 29 inch (736mm) nominal diameter hydraulic coupling. The two 35HP motors drive a 10ft 3-inch (3.12m) diameter fan with a capacity of 166,000 cubic feet of air per minute through a 23 inch (584mm) nominal diameter hydraulic coupling and through single reduction gear.
Taylor	1 No	1 No

Street	30HP and 735 rpm 9HP and 485 rpm 1 No 30Hp and 735 rpm 4HP and 365rpm Both motors drive a 9ft (2.74m) diameter fan, with a capacity of 137,000 cubic feet of air per minute through 23 inch (584mm) nominal diameter hydraulic couplings and bevel single reduction gear	40HP and 735 rpm 12.5HP and 485 rpm 1 No 40HP and 735 rpm 6HP and 365 rpm Both motors drive a 10 ft 3-inch (3.12m) diameter fan with a capacity of 4173,000 cubic feet of air per minute through 23 inch (584mm) nominal hydraulic coupling and bevel single reduction gear

Queensway Ventilation Diagram (Wirral Archives)

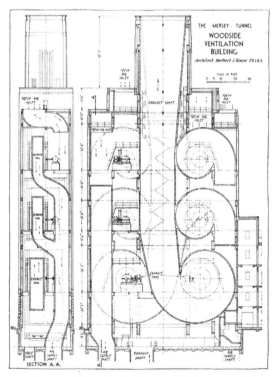

Section through Ventilation Tower (Birkenhead Library)

Construction of Tunnel Ventilation Shafts (Birkenhead Library

Bretherton Buildings Axis Supply Duct (Birkenhead Library)

Brick Assembly to Supply Shaft (Birkenhead Library)

Exhaust Chamber and Tunnel Roof (Liverpool Museum)

Fresh Air Ducts Land Side (Liverpool Museum)

Fresh Air Ducts Construction (Liverpool Museum)

Supply Duct Shuttering (Liverpool Museum)

Walker Fan Blade (Liverpool Museum)

Sturdivant Fan Impeller (Liverpool Museum)

Self-Moving Fan (St Georges Dock Building) (Author)

St Georges Dock Building from the Strand (Author)

St Georges Dock Fan (Liverpool Museum)

New Quay Ventilation Construction (Liverpool Museum)

Liverpool New Quay (Author)

North John Street under Construction (Liverpool Museum)

North John Street Today (Author)

Sidney Street under Construction (Liverpool Museum)

Sidney Street Ventilation Station (Author)

Taylor Street Ventilation Station (Author)

Woodside Ventilation Station (Author)

Wirral Ventilation Buildings from Pier Head (Author)
This view is normally obscured by the Irish ferry at the terminal shown by the five white tipped posts

Original Morpeth Dock Pump Room (Liverpool Museum)

Original mid River Pump Room (Liverpool Museum)

Chapter 10
Finishes

Bituminous Paint

In the early days of the tunnels inception, thought was given to the gasses and fumes given off by the vehicles in the tunnel during its construction and daily use. The engineers decided to conduct tests on the available materials to see which one worked better in withstanding the atmospheric conditions. They first looked at the traffic that used the ferries along with the composition of the gasses and its possible effects on the tunnel as the vehicles passed through.

To enable the process of finding the composition, special apparatus was designed to produce a mixture of these gases, carbon dioxide, carbon monoxide, sulphur dioxide, and air saturated with water. The mixtures were fed to tanks kept at a temperature of 6 degrees, and specimens of steel, cast iron, aluminium and copper alloys, cements, concretes and particularly paints, were placed in the tanks. Some of these materials were quickly affected by the gases, but a few stood the test.

Among these were the products of the Bituminous Compositions Ltd of Grimsby. The company's paints, spread on cast iron, steel cement and concrete, were exposed to the gases over a period of three years. They stood up to the rigorous conditions so well that they were adopted for the painting and preserving of a vast quantity of the material used.

Asphalt, also known as bitumen, is a sticky, black and highly viscous liquid or semi-solid form of petroleum. It may be found in natural deposits or may be a refined product; usually referred to as pitch. The primary use of bitumen is in road construction, and its other main use is for various waterproofing products. Naturally occurring asphalt/bitumen is sometimes specified by the term 'crude bitumen'. Its viscosity is similar to that of cold molasses while the material obtained from the fractional distillation of crude oil, which has a boiling point at 525 °C (977 °F).

One of the earlier uses of Asphalt in the United Kingdom was for etching. William Salmon's Polygraphice (1673) provides a recipe for varnish used in etching, consisting of three ounces of virgin wax, two ounces of mastic, and one ounce of asphaltum. By the fifth edition in 1685, he had included more asphaltum recipes from other sources. The first British patent for the use of asphalt/bitumen was Cassell's patent asphalte or bitumen' in 1834. Then on 25 November 1837, Richard Tappin Claridge patented the use of Seyssel asphalt (patent #7849), for use in asphalte pavement.

Trials were made in 1838 on the footway in Whitehall, the stable at Knightsbridge Barracks, and subsequently on the space at the bottom of the steps leading from Waterloo Place to St. James Park. The formation in 1838 of Claridge's Patent Asphalte Company gave an enormous impetus to the development of a British asphalt industry. By the end of 1838, at least two other companies, Robinson's and the Bastenne company, were in production, with asphalt being laid as paving at Brighton, Herne

Bay, Canterbury, Kensington, the Strand, and a large floor area in Bunhill-row, while meantime Claridge's Whitehall paving continued in good order.

In 1838, there was a flurry of entrepreneurial activity over asphalt/bitumen, which had uses beyond paving. For example, asphalt could also be used for flooring, damp proofing in buildings, and for waterproofing of various types of pools and baths.

Gunite Rendering

Gunite is a method of rendering by the use of a cement gun system which is advantageous for work within a tunnel situation. The materials are sand and cement, which are first mixed together in the dry, and then loaded into the cement gun. From the gun, the materials are carried by compressed air through a flexible hose to the nozzle. From here, the material is discharged in a steady stream and under considerable pressure on to the surface to be rendered. Water necessary for hydration is introduced at the ejecting nozzle, where it is fed into the stream of materials in the form of a finely divided spray, and is thus intimately mixed with the cement.

The system allows the user to build up the layers of concrete, and a perfect bond to the surface. It is for this reason that sprayed concrete is widely used for various projects. Before 1926, the cement gun system had been used rather extensively in in the construction of tunnels in the United States, France, Germany and elsewhere, but only to a relatively small extent in Great Britain. The Queensway Tunnel engineers began investigating the use of a cement gun to solve the problems associated with the confined spaces within the Tunnel.

The cement gun played its part in the construction after its introduction in 1927, and its continuous use until the completion of the main structure in 1933. This work was carried out by the Cement-Gun Co Ltd as sub-contractors to Nuttalls. The gunite covered 12,700 superficial yards, (11.61 metres) and during the early part of 1931, a full-size section of reinforced gunite mantle lining in the roadway arch was constructed for experimental purposes. This was left several months for observation and test.

At about the same time, the Cement-Gun Co Ltd carried out several small works in the tunnel, both reinforced and un-reinforced portions of the North John Street access passage arch, portions of the arch and side walls of the New Quay section of the tunnel, some cable shafts, and the waterproofing of the mid-river pump-room.

In view of the success demonstrated by the foregoing work, the use of gunite was considered by the engineers for lining the roadway arch, the requirements being a smooth and watertight covering to the concrete already placed between the cast-iron supporting rings.

In order to investigate thoroughly all details, a full-size section of roadway arch lining was constructed in the autumn of 1932, under the direction of the engineers. The results of tests on this section eventually led to the covering of the whole of the roadway arch, above the glass dados, with reinforced gunite. This work was carried out by the Concrete Proofing Co Ltd of London, as sub-contractors to both Nuttalls' and Messrs McAlpines'.

The first step in the carrying out of the work was to thoroughly clean the existing surface to which the gunite lining was to be applied. This was done by hosing down with area; although, the use of wire brushes and pneumatic scaling hammers was sometimes found necessary. In addition, any smooth areas of the concrete archway were hacked over to assist in providing bond. On completion of this preliminary work the reinforcement was fixed in position, it being kept tin clear of the concrete backing by inserting small distance-pieces. This was to ensure that the rendering would thoroughly incorporate the reinforcement, which was thus placed centrally in it.

For the purpose of smoothing off the gunite rendering to a true surface, and also to obtain the specified thickness, light wooden strips or screeds were next fixed in position round the perimeter of the arch on top of the reinforcement. They were placed at 8 feet (2.43 metres) intervals and fixed to the concrete by means of rawlplugs and screws.

The gunite rendering was now applied to its full thickness in one application, each bay (the 8 ft wide panel enclosed between screeds) was completed at a time. When two bays had been finished the intermediate screed was removed, and the space previously occupied by the screed filled in. Light timber gantries from which the entire perimeter of the arch was easily accessible were used for scaffolding. Mounted on wheels, their lightness and mobility greatly facilitated the progress of the work. They were easily convertible for use in either the 44 ft (13.4 metres) wide tunnel or the 26 ft 6in (8.1 metres) tunnels, and later in the junction chambers and ventilation and exit shafts.

Compressed air for operating the cement guns was supplied by the main contractors from each side of the river, and lighting was provided by 500-watt portable floodlights mounted on trestles. The work commenced in May 1932, and proceeded rapidly with six cement guns used continuously on both the day and night shifts. The greater part of the work on the roadway arch was completed by October 1932, but additional works authorised by the engineers from time to time substantially increased the value of the contract. These additional works consisted of reinforced gunite rendering to the junction chambers and several of the shafts and adits of the ventilation and emergency exit systems. The entire gunite work was eventually completed in the early summer of 1933.

Altogether, the superficial area of the gunite placed in the Queensway Tunnel exceeds 20 acres, and varies in thickness from ¼ to 4 inches (6 mm to 100 mm). The use of such a large amount of the material in a structure of such universal engineering importance would seem entirely to obviate discussion as to the sufficiency and reliability of the material, especially in view of the very careful investigations, tests and trials made by the engineers

Architectural Features

The Queensway Tunnel Architect (Mr Rowse) was not only responsible for the six ventilation stations, but he was also responsible for the four entrances, and approach portals along with toll booths, walls, lamps, glass lining and many more features both seen and unseen in the tunnel.

The main entrances have arched pylons that are more than being a decorative feature. They are utilitarian and the aesthetic are everywhere closely allied in the visible parts of the tunnel workings. Besides marking the limits of the approaches to oncoming traffic, these arches contain accommodation for the staff employed in the tollbooths that lie across the entrances.

The original tollbooths are of cast iron with a dado of decorative metalwork, the upper portion is in glass with opening sections, and lighting cove all round. Above the windows, there is a symbolic design suggestive of speed. These booths were emerald green and gold. Dominating the whole area about the main entrances at Liverpool and Birkenhead are the great lighting shafts, 60 ft (18.28 metres) high, constructed of reinforced concrete and overlaid with fluted and polished black granite. The shafts are surmounted by glazed bowls of gilded bronze with a decorative pinnacle above, serving to mark the entrance for approaching traffic. The idea was to provide adequate lighting for the area round the entrances, and at the same time to give some monumental expression to this great engineering achievement. The hollow columns contain ladders to provide easy access for maintenance purposes.

The Liverpool shaft was erected in reinforced concrete by Natal, and the Birkenhead shaft by McAlpine and Sons. The Trussed Concrete Steel Co supplied the steelwork, whilst the fabrication and erection of the granite was carried out by John Stubbs & Sons. The shafts are not merely tapered, but are slightly curved to correct the optical illusion of being a smaller diameter in the centre than at the top of the shaft. Every section of the granite was shaped individually to the change in the curve of the shaft. Bronze tablets have been attached at the base of each lighting shaft with the centre tablet commemorating the Royal opening of the tunnel by His Majesty the King. Another records the names of the members and officers of the Mersey Tunnel Joint Committee. A third tablet incorporates the names of the engineers, the architect and the valuer, whilst the fourth is an acknowledgment of the services performed by the principal contractors.

The Portland Stone portals are surmounted by immense carvings designed by Thompson & Capstick, in association with the architect. Over the Liverpool entrance, an ornamented circular shield supported by two flying bulls of strikingly vital conception, and with a winged wheel above, all symbolic of swift and heavy traffic. The sculpture for the Birkenhead entrance is more formal, but no less striking. In this case, a triangular shield is used, over which there appears the head of a figure suggestive of a motorist at speed.

Within the tunnel, itself, nobody can deny the beauty of the black glass lining which ran the full length of the tunnel (now covered by the modern white panelling). The design and the finished appearance were equally simple and easily cleaned, as well as being easy to remove without difficulty for repairs or replacement.

By way of contrast to the run of the glass dado, the fire stations create effective points of colour; they are in brilliantly illuminated red glass. Above each station is an illuminated sign, which throws a shaft of light down on to it, and, when so controlled from the central operation room, can signal instructions to drivers of vehicles. The railing along the footpath at the sides of the tunnel (now removed after the installation of the safety areas protruding from the tunnel walls, were originally painted red to clearly denote the sides of the roadway.

The lamp standards at the New Quay and Birkenhead entrances are made of special centrifugally spun concrete with a white Portland stone finish. These were manufactured by the Liverpool Artificial Stone Co Ltd.

In addition to the lamp standards, an immense quantity of paving slabs and kerbing were also supplied by the same company who installed special plant to meet the requirements of the engineers. The paving slabs were manufactured with British granite and cement, under hydraulic pressure of 600 tons. They had to be reinforced to withstand a specified load and given a special carborundum stone finish, because they cover the conduits carrying power and telephone cables through the tunnel.

The ornamental cast-iron work to the central lighting shafts, the tollbooths, and the parapet and wall lamps was carried out by Messrs. H. H. Martyn, whose work throughout displays the finest qualities of metal craftsmanship. The interior steel linings and equipment of the tollbooths have been supplied by Messrs. Roneo Ltd.

The electrical equipment of the tollbooths, the central lighting shafts and the parapet and wall lamps was undertaken by Higgins & Griffiths. The electrical clocks and calendars in the tollbooths were supplied by the Synchromatic Time Recording Co Ltd.

Polished Granit Podium

This is a feature that has proved an interesting one to visitors to Liverpool. It shows the traverse of the tunnel itself under the riverbed, branching out into its four entrances, and also the positions and characteristic profiles of the six ventilation buildings, three on each side of the river.

The Glass Dado

When the final coat of 'Marplax' finish had been applied to the walls above the dado line, it was the choice of glass, and, both Mellowes Ltd of Sheffield, and Pilkington Ltd of St Helens played a considerable part. It was simple in design and of elegant appearance and can be easily cleaned and the maintenance costs are practically nil.

The dado is made up of ¼-inch (6 mm) thick sheets of black glass, laid in lead caves and framed with stainless-steel bars. The natural vitreous surface of the glass provides a perfectly durable lining, which is unaffected by moisture or fumes. The dado runs the full length of the tunnel to a height of 6 feet 3 inches (1.9 metres) from the concrete plinth level, and consists of three graduated rows of glass in large sheets supported by means of special rails of non-ferrous metal. These rails are designed with a movable wing or web at the front and with two contact cushions at the back, so that whilst the glass is held firmly in position, there remains sufficient room for slight movement due to contraction, expansion or vibration. This member is also utilised for the vertical joints, which are at approximately 10-foot centres (3.04 metres).

The rails are in turn supported by special screws driven into solid lead plugs inserted into concrete facing and driven home by pressure to a minimum depth of two inches (50 mm) and left protruding from the face of the concrete by 5/8 of an inch (15.6 mm).

This is essential to provide and preserve an insulating space between the concrete and the back of the glass, so as to allow variation of alignment to be overcome, and also to give accommodation for lighting cables, condensation tubes, etc., and to prohibit any movement of the main structure being transferred to the surface of the glass.

The screws are of stainless steel, with a specially designed thread, and of a length of the whole of the thread is embedded in the plug

These supporting members, both vertical and horizontal, are ultimately faced by special stainless-steel members which added to the security and support of the glass, and are of a profile which prevents the lodgement of dust and facilitates cleaning. Special treatment has been given to the doorways, splays, junctions, etc., all of which conform to the general alignment and design.

The Plazas

At each of the four entrances were planned by the tunnel engineers with the co-operation of the architect, and in consultation, on the Liverpool entrances with the City Engineer, and at Birkenhead, the Borough Engineer. Their construction has involved a considerable amount of thought as to their use and ease of allowing traffic to flow freely in and out of the tunnels. The scale and character of the plazas are worthy of the great engineering feat which the tunnel represents.

Paving and Tramway Junctions

The layout of the tunnel entrances in Liverpool originally necessitated the entire reconstruction of the street paving to different levels over an area of approximately 21 acres, and adjacent to these entrances, at a cost of £59,000 (£3,678,667.70). The department responsible for this was also responsible for co-ordinating the work of the various public service authorities, whose service areas covered approximately 1 ¼ miles in length, affected the street works.

It was also necessary to re-set the location of important tramway junctions and tracks, which link up five main routes at the Old Haymarket. This re-arrangement necessitated re-designing the whole of the junctions and has involved 1,500 linear yards (1371 Metres) of new tramway track, 30 points and 70 crossings.

This work was carried out during traffic hours at a cost of approximately £20,000 (£1,247,006.00). The public underground toilets in the Old Haymarket had to be demolished and filled in. New toilets were concealed behind the boundary walls of St. John's Gardens, which were erected in St. John's Lane at an approximate cost of £7,500 (£467,627.25).

The Concluding Stages (1933–1934)

Work on the tunnel progressed at a steady rate until June 1934 and the additional works were to become the six new ventilation stations which cost in excess of £570,000 (£35,539,671.00). These presented a problem, not only for the tunnel and its eventual cost, but firstly, how to solve the ventilation problem and secondly, how to build such aesthetically pleasing structures in a style befitting the Liverpool Skyline and heritage.

Time was not on their side nor was the solution as the dates were set for the tunnel opening and the work on the tunnel was progressing on. In other areas of the tunnel work, the surfacing of the roadways with studded iron plates, the ornamental glass dado, five feet high, and the finishing touches to the great arches with cement and fine plaster.

At Christmas 1933, and Easter 1934, the tunnel was thrown open to the public, in the cause of charity, and nearly 300,000 people travelled through the main tunnel on foot, and expressed their admiration for an achievement that will forever be a credit to British science and craftsmanship.

As the public were getting their view of the tunnel for the first time, the large traffic areas at the Old Haymarket, Liverpool, and Chester Street, Birkenhead as well as the two dock-side entrances, were under construction. Although, construction may seem a progressive word, the reality of this project was that a large number of properties and businesses were moved to other locations to make way for the tunnel. The result was a large-scale movement of people and the eventual destruction of the properties they once occupied. In addition to the works, more improvements were being undertaken on the periphery of the tunnel works in both areas which were being improved and additional lighting added. The lighting would include 60 feet (18.28 metres) high lighting columns that stand sentinel at the tunnel mouths.

Now the question of how the tolls were to be collected and all the little points that go to make up a large project such as the Queensway Tunnel. The tunnels would have to have its own byelaws to allow it to operate and administer itself. To enable them to see how other tunnels operate, the tunnel committee visited the Scheldt tunnel at Antwerp. The committee decided to enter into contracts with the Corporations of Liverpool and Birkenhead to take over the various services legal, accountancy, engineering, cleansing, policing, and breakdown and this system would result in a saving of around £20,000 (£1,247,006.00) a year, as compared with the cost of a full-time staff. Mr F. Robinson, Assistant Manager of the Woodside Ferries, was appointed as Traffic Manager for the tunnel.

Proposed tolls and by-laws have been submitted to public criticism, and amended, but were regarded as experimental. For the initial period, pedal cycles and steam vehicles of an approved type were admitted, but horse-drawn and hand-propelled vehicles, pedestrians, and animals on foot were excluded. The Tunnel Committee reserved to the right to deny admittance to any vehicles not conforming to regulations.

Water Supply

The supply of water required for the cleansing of the tunnel and the service of the hydrants situated at each of the fire stations has necessitated the installation of a 4-inch (100 mm) main, which runs throughout the entire length, and is tapped at convenient points so that water can be drawn as required. At the boundary line between Liverpool and Birkenhead, a stop-valve has been placed.

In the ordinary way this valve will remain closed. The Liverpool and Birkenhead sections of the tunnels are supplied with water by the respective Corporations. In the remote event of a failure of water supply at either end, the stop-valve will be opened to enable the whole length of the main to be fully supplied from the alternative supply.

Casting Moulds for the Iron Road Sections (Birkenhead Library)

Spraying the chills with sulphur.

Note the lack of breathing apparatus to protect from the fumes of the sulphur spray onto the moulds (Birkenhead Library)

Casting the Iron Road Sections (Birkenhead Library)

Laying the Iron Road Sections (Birkenhead Library)

Birkenhead Opening Ceremony Commemorative Column and Plaques, circa 1930s
(Birkenhead Library)

Opening Ceremony Commemorative Column and Plaques Today (Author)

Opening Ceremony Commemorative Column Plaques (Author)

Opening Ceremony Commemorative Column Plaques (Author)

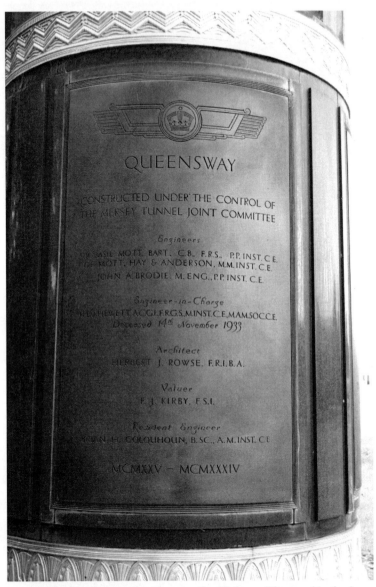

Opening Ceremony Commemorative Column Plaques (Author)

Opening Ceremony Commemorative Column Plaques (Author)

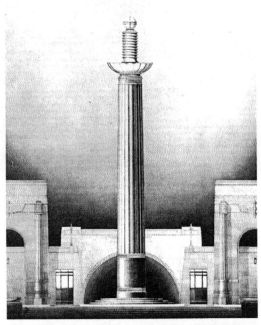

Liverpool Entrance Tunnel Column Drawing (Birkenhead Library)

Birkenhead Tunnel Approach circa 1934 (Birkenhead Library)

Early Days of Tunnel in Use (Birkenhead Library)

Liverpool (Haymarket) Entrance (Liverpool Museum)

Dock Exit Today (Author)

Rendel Street Entrance (Author)

Lower Section of Rendel Street to Actual Tunnel Entrance (Author)

View out of Rendel Street onto Approach Road.
A view very few will have seen in the past (Author)

Rendel Street Tunnel (Author)

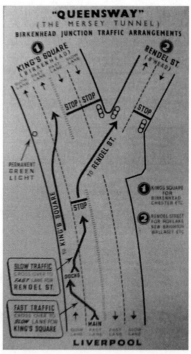

Tunnel Direction for use of Rendel Street (Birkenhead Library)
(Note two-way traffic in and out crossing live tunnel lanes, which would not be allowed today) (Wirral Archives)

Chapter 11
Queensway Tunnel Opening Ceremony

This thoroughfare is great and strange. The wonder of your Tunnel will only come into mind after reflection. Who can reflect without awe that the will and power of men, which in our time have created the noble bridges of the Thames, the Forth and Sydney Harbour, can drive also tunnels such as this, in which many streams of wheeled traffic may run in light and safety below the depths and turbulence of tidal water bearing ships of the world?

"*Many hundreds have toiled here and the work of many thousands all over the country has helped their toil. I thank all those whose effort has achieved this miracle.*

"*May those who use it ever keep grateful thought of the many who struggled for long months against mud and darkness to bring it into being. May our people always work together thus for the blessing of this Kingdom by wise and noble uses of power that man has won from nature.*"

King George V–18th July 1934

With these words the King, watched by 150–200,000 people, activated the gold switch to raise the curtains covering the entrance to the Tunnel and officially opening one of the greatest feats of British engineering to the public. An undertaking which has been years in the planning and 9 years in construction was finally complete.

An article in the local Liverpool Echo the day before the opening, had was a piece titled *The Old Salts Lament 'New Fangled Things – Like Tunnels'*. The reporter was talking to an old seaman who had worked on the river and at sea for all his life and was not too impressed with the 'new-fangled tunnels'.

He questioned the reporter by asking in his husky almost angry voice,

"I s'pose you don't see no beauty in that. I s'pose the lies of you sees only beauty in tunnels and them new-fangled tunnels," as they sat on the railings of the ship he had just brought into Liverpool.

Waving his arm up and down the river he further stated, "The whole lots picturesque, nothing like it anywhere and I've spent my whole life on it, I 'ave."

Further on in the discussion, he noted, "Them as goes motoring through this ere tunnel won't even know they have been in Liverpool. They won't! All they know is they went down a hole in one place and came out of a hole in another. This here river's Liverpool! If a town has got what they call a soul, this is it (pointing to the river) down here. St Georges Hall isn't Liverpool, The town hall isn't Liverpool and the streets isn't either. You go around the world and every port you come to you'll see 'Liverpool' on some ship."

Commenting on his using the tunnel the old seaman stated,

"Not if the Lord Mayor and his own self goes on his knees to me, I won't go through it. Because, them what wants to burrow can burrow, but whilst I've my eyes and that's things to see on top, they won't get me to be no bloomin' rabbit."

Pre-Celebration Pageant the evening before the opening Ceremony:
The night before the opening ceremony, there were a series of evening concerts in Exchange Flags a few minutes' walk up Dale Street towards the river and St Johns Gardens (opposite the tunnel entrance):

- 5 p.m. Exchange Flags – Liverpool Civic Orchestra
- 7 p.m. Band of the Grenadier Guards leading procession
- 7:30 p.m. St Johns Gardens – Liverpool City Tramways Band
- 7:30 p.m. Exchange Flags – Liverpool City Police Band

Route of the procession (changed from the original advertised route): Assemble in Victoria Street and move off to North Johns Street, Dale Street, Castle Street, Lord Street, Church Street, Lord Street, North john Street, Victoria Street where the procession will disperse.

There was no view of the pageant procession from the Haymarket sites (tunnel entrance) or from St Johns gardens (Opposite the entrance and behind St George's Hall).

The band of the Grenadier Guards were at the front of the procession followed by The Spirit of Liverpool, The Herald, Commerce, Faith, Remembrance, Charity, Vision, Progress and Youth. Following on from this there were a number of vehicles, each of which had its own symbolic meaning.

Banking – with a figure of Britannia riding aloft

Insurance – representing the different services of insurance. Salvage car maintained by the insurance companies in attendance.

Railways – the power of rail transport, with accompanying horses of the companies.

Mersey Docks and Harbour Boards – The services rendered by the dock board and the life and beauty of the sea.

Shipping – the wide spread of Liverpool shipping accompanied by various members of a ships company and representatives of the different countries with which Liverpool trades.

David Lewis band

Communication Car – the powerful interest of the telegraph of communications and the post office.

Newspaper – showing the mightiness of the pen

Cotton Car – the bulk and majesty of the cotton plants and bales with colourful representatives of finished cotton

Milling – the great milling industry of Liverpool

Provisions – the provisions market supporting and surrounding the breakfast table

Fruit – A mass of fruit crowned by the goodness who symbolises fruit with girls carrying various types of fruit.

Timber – The huge logs that serve the manifold uses to which they are put.

Gas – Exhibit showing the great power and service of gas represented symbolically

Pageant Master – Mr Harold king

Stewards – Miss N. E. Law, Mr George Creed, and Mr W. J. Green

8 p.m. – Pageant Plat at Exchanged Flags.

Opening Ceremony

The day's programme of events was full and in order to do the day's monumental events justice I have listed below the events not only for the opening of the tunnel but the opening of the new main library in Birkenhead which replaced the Carnegie Library demolished to make way for the new tunnel.

10:50

Their Majesties the King and Queen arrived at the city boundary, East Lancashire Road and were met by the Lord Mayor and the Lady Mayoress (Councillor and Mrs George A Strong) and the Lord Mayor presented the Lady Mayoress, Sir Thomas white, the Town Clerk (Mr Walter Moon), and the Chief Constable (Mr A K Wilson). Their Majesties the King and Queen left in the Royal car escorted by the motor Police. We would know this better as the Police motorcycle outriders on escort duty. The procession proceeded along East Lancashire road and onto Walton Hall Park in the following order.

(The following has been taken from local newspapers at the time and it is unclear if the day's events timings were strictly adhered to)
Private car
The Earl and Countess of Derby
Miss Ruth Primrose
Lord Stanley

Royal Car
THEIR MAJESTIES THE KING AND QUEEN
Dowager, Countess of Airlie (Lady in Waiting)
(The Countess was to serve Queen Victoria, Queen Alexandria and Queen Mary before she died in 1956)
Mr Leslie Hore-Belsha M. P. (Minister of Attendance)

Private Car
Lady Sefton
Lady Maureen Stanley
Lord Sefton
Major Hon. A Hardinge (Assistant Private Secretary)

Private car
Captain bullock
Lieutenant Colonel R. H. Seymour (Equerry in Waiting)

11:10

Their Majesties, the King and Queen, were received at Walton Hall Park by the Chairman of the Parks and Gardens Committee (Mr George Holme), who was presented by the Earl of Derby. The following were then presented, Councillor Peter Kavanagh, Deputy Chairman of the Parks and Gardens Committee, and Mr J. J. Guttridge who was the chief superintendent and curator of the parks and gardens.

Walton Hall Park had been extended at a cost of £37,000 (£2,362,174.87) to include a spacious boating lake and boathouse for 48 boats, a two-acre model boat lake, two storeys open air café with verandas on the ground and spacious balconied above, all overlooking the lake, a bandstand, boulevards, walks, cricket tennis, football and baseball grounds and a children's open air gymnasium, for adults there were the putting greens, bowling greens and tennis courts. The auditorium which fronts the bandstand can now be enlarged to cover a capacity of 15,000 people and used for open air plays, pageants, concerts and other presentations.

His Majesty the King duly declared Walton Hall Park open without leaving the car.

11:15

The Royal procession proceeded via Walton Hall Avenue, Queens Drive, Muirhead Avenue, Derby Road, through Newsham Park by way of the Seamans Orphanage to Prescot Road, Kensington, Prescott Street, London Road and William Brown Street and the length of Kingsway to the entrance of the Tunnel.

As the Royal Party made their way down London Road, the crown of 150–200,000 people joined in to sing the first verse of Elgars 'Land of Hope and Glory' accompanied by the band of the Liverpool City Police. The singing turned to murmuring as everyone was standing on tiptoe to get a view of the King and Queen as they went past. Some of the crowd had prepared themselves for a long day and had brought provisions for at least one meal and it was said that the scenes were reminiscent of Armistice Day. The most striking form of 'decoration' was that formed by a group of over a thousand school children who, dressed in vivid coloured frocks and hats and carefully grouped on the steps of museum in William Brown Street. The sight of this display formed a human floral bouquet which tier upon tier on the museum steps. This was all part of the planned day's events

11:45

A portion of the initial route to the tunnel was lined by the Royal Naval Volunteer Reserve (Mersey Division) and the 55th (West Lancashire) Division (Liverpool) along with the British Legion (Now Royal British Legion).

Their Majesties the King and Queen were received by the Lord Mayor and the Chairman of the Mersey Tunnel Joint Committee (Sir Thomas White). Lord Derby as Lord Lieutenant of Lancashire presented The General Officer Commanding Western Command (Lt General Sir Walter M. St. G. Kirke) and the General Officer Commanding West Lancashire Area and 55th (West Lancashire) Division (Major General W. J. N. Cooke-Collins).

His Majesty the King inspected the Guard of Honour mounted by the 2nd Battalion, The Kings Regiment (Liverpool) and simultaneously a detachment of the West Lancashire Territorial Army Nursing Service was inspected by her Majesty the Queen accompanied by the Countess of Derby and the Lady Mayoress.

A procession was then formed as follows whilst the Royal Party and dignitaries make their way to the Dais for the official opening ceremony.

Civic Regalia
The Town Clerk, Sir Thomas White
The Lord Mayor
THEIR MAJESTIES THE KING AND QUEEN
The Minister in Attendance
(Mr Leslie Hore-Belisha, M. P.)
Lord Derby
The Ladies and Gentlemen in Waiting

Upon their Majesties taking their seats, the Lord Mayor tendered the city's welcome and asks their Majesties gracious acceptance of an address from the Mersey Tunnel Joint Committee, Sir Thomas White (Chairman of the Mersey Tunnels Joint Committee) read the address and handed the address to his Majesty the King upon completion.

Sir Thomas White read:

"To the King's most excellent Majesty
"MAY IT PLEASE YOUR MAJESTY

"We your loyal subjects, the Members of the Mersey Tunnel Joint Committee beg leave to approach your Majesty and her Majesty the Queen with an expression of our dutiful homage and our gratification at the high privilege of welcoming your Majesties on this the memorable occasion of your visit to Liverpool and Birkenhead, and we assure your majesties of our deepest loyalty and affection.

"It is a matter of great gratification to the citizens and burgesses that your majesties have once more evinced deep interest in all that concerns the welfare of your people by your presence here today for the purpose of opening to public traffic the largest sub aqueous tunnel in the world – a great engineering work of natural character and importance which has been constructed to meet the ever growing transport requirements of the Port of Liverpool and an event of great importance to our trade and local history.

"You Majesties presence on this eventful occasion is regarded by the citizens and burgesses of Liverpool and Birkenhead as another example of the profound interest taken by your Majesty in all great projects having for their object the welfare and prosperity of your Majesties subjects.

"It is our heartfelt prayer that the almighty may continue to bestow every blessing upon your majesties and that your Majesties may both be long spared to this great empire to advance the happiness to which your best endeavours are always directed.

"Given under the common seal of the Mersey Tunnel Joint Committee this eighteenth day of July One Thousand and Thirty-Four.

His Majesty the King read his reply and officially declared the tunnel,

"Thank you for your address to the Queen and myself. It is with deep pleasure for us to come here today to open for the use of men a thoroughfare so great and strange as this Mersey Tunnel now made ready by your labour.
"In some other seaport channels and estuaries have been bridged with structures which rank among the wonders of the world. Such bridges stand in the light to be

marvelled at by all. The wonder of your tunnel will only come into the mind after reflection.

"Who can reflect without awe that will and power of men, which in our own time can create the noble bridge of the Thames, the fourth, the Hudson and Sydney harbour, can drive also tunnels such as this, wherein many streams of wheeled traffic may run. In light and safety below the depth and turbulence of a tidal water bearing the ships of the world?

"Such a task can only be achieved by the endeavours of a multitude. Hundreds have toiled here, the work of many thousands all over the country has helped their toil. I thank all those who efforts have achieved this miracle.

"I praise the imagination that foresaw the minds that planned, the skill that fashioned, the will that drove, and the strong arms that endures in the bringing of this work to completion.

"May your peoples always work together thus for the blessing of this kingdom by wise and noble uses of the power than man has won from nature.

"I trust that the citizens of this double city, so long famous as daring traders and matchless seamen, may for many generations find profit and comfort in this link that binds them.

"I am happy to declare the Mersey Tunnel open. May those who use it ever keep grateful thought of the many who struggled for long months against mud and darkness to bring it into being.

Whilst the King was giving his response, Sir Thomas White and the people of both Liverpool and Birkenhead, listened intently at the King's reply through the various speakers positioned around the areas.

Declaring 'Queensway' open by operating an electrical switch which simultaneously indicated at the Kings Square Entrance, Birkenhead that the tunnel was now officially open. His majesty's speech was broadcast and also repeated by speakers on both sides of the river. At the close of the address, his Majesty pressed the switch which raised the two yellow poles (one either side of the entrance to the tunnel) and the emerald green curtains, which had draped across the tunnel entrance parted and each half of the tunnel entrance was revealed.

As the curtains raised, they revealed a message positioned across the entrance and in which red letters spelled out 'Merseyside Welcomes Your Majesties' along with the blackness of the mystical tunnel entrance.

As the national anthem played and the curtains began to rise, few were aware that the electrical mechanism had failed and instead two men were stationed either side, raising the curtains with hand cranks. They did such a good job of opening the curtains in synchronisation that no one noticed. I have tried to find the names of these men to give them the credit they deserve.

To commemorate the event, 150,000 local children were awarded medals and the city held a week of celebrations which included a Ceremony of Remembrance at the Cenotaph at St Georges Hall (Opposite Liverpool Lime Street Station) in the city centre.

At the time of its construction, the Queensway Tunnel was the longest underwater tunnel in the world, a title it held for 24 years when the Kanmon Strait Tunnel, 200 metres longer, was built in Japan.

The Chairman of the Mersey Tunnels Joint Committee requested his Majesty the King if he would graciously accept as a moment of the occasion, a small model of the Tunnel Entrance, which his Majesty had used when declaring the tunnel open.

He also asked her Majesty the Queen if she would graciously accept a Georgian Silver Cup. The Police Band would then play the hymn *All people that on earth do dwell* and the Lord Bishop of Chester offered The Lord's Prayer, followed by the Lord Bishop of Liverpool offering the Benediction.

The Chairman of the Mersey Tunnels Joint Committee (Sir Thomas White) presented the following members of the Mersey Tunnel Joint Committee:

- Alderman Denis J Clarke Deputy Chairman
- Alderman John Clancy
- Alderman Thomas Dowd
- Alderman Burton William Ellis
- Alderman Austin Harford
- Alderman Luke Logan M.B.E.
- Alderman William Muirhead
- Alderman William albert Robinson
- Alderman Edwin Thompson
- Councillor Robert John Hall
- Alderman Thomas McLellan
- Alderman Godfrey Allan Solly
- Alderman Henry Van Gruisen
- Councillor William Henry Egan
- Councillor Edward James Hughes
- Councillor Charles McVey
- Councillor Alfred Gates (former Lord Mayor of Liverpool)
- Sir John Sandeman Allen (senior Member of Parliament for Liverpool)
- Sir Thomas White then presented
- The Chief Engineer, Sir Basil Mott Bart C. B.
- The Joint Engineer, Mr John A Brodie
- The treasurer, Mr W. H. Legh Smith
- The Architect, Mr Herbert Rowse
- The Valuer, Mr Francis J Kirby
- The town Clerk of Birkenhead Mr E. W. Tame
- The City Engineer of Liverpool, Mr T Molyneux
- The City Electrical engineer of Liverpool, Mr P. J. Robinson
- Sir Thomas White then presented one representative from each of the Contractors
- Mr G. G. Lynde representing Messers Edmund Nuttall Sons and Co Ltd
- Sir Alfred McAlpine, representing Messers Sir Robert McAlpine and Sons
- Mr J. A. Strange, representing Messers William moss and Sons Ltd
- Mr Charles Boot, representing Messers henry Boot and Sons Ltd
- Mr George M Burt representing Messers John Mowlem and Co Ltd
- Sir Felix Pole, representing Messers Metropolitan Vickers Electrical Company Ltd
- Sir Thomas white then presented a foreman workman from each of the six main contractors

Following on from this, his Majesty presented his gold medal to the Conway Cadet (Harold Charles Kirby).

The president (Captain J. A. Coverley) and the captain of the Conway (Commander M. G. Douglas R.D. R.N.R.) was presented to his Majesty the King.

12:15

The King and Queen left the Dais to enter the Royal Car and the first verse of the National Anthem was sung led by the band of the 2nd Battalion the Kings Regiment (Liverpool). Their Majesties then drove through the tunnel to Kings Square, Birkenhead and stopped at the dais in front of the stand.

After the King and Queen had entered the tunnel, the dignitaries went off to the Liverpool Town Hall for a reception.

12:25

After greeting some of the people on the dais, Lord Derby walked across the intervening space towards the Grenadier Guards and shook hands with Captain Miller, the conductor who left the stand to greet his lordship. They both stood chatting for a few seconds. Lord Derby was responsible for the day's events and he had undertaken to request Captain Miller to oversee the events on the Birkenhead side.

With the royal procession expected, at any moment, one of the Boy scouts, who acted as a guard of honour in the tunnel entrance, fainted, and he was quickly carried away by his colleagues to receive first aid from one of the many first aid post positioned around the area.

When the royal car was being driven up the incline to the Birkenhead exit, the territorial unit lowered their colours (unit flag depicting battle honours) and at the same time the Royal Standard was broken (opened and displayed) at the head of the slender flagpole immediately above the tunnel cutting.

This was a signal for the cheering crowd which was begun by school children who had the highest and best view of the das events.

The Earl of Derby stepped off the dais to greet the royal party presented:
- The Mayor of Birkenhead (Councillor James Coulthard J.P.)
- The Mayoress (Mrs R. L. Jones daughter of the Mayor)

Following from this the Mayor of Birkenhead presented:
- The Recorder (Mr Clive T Wilson M. P.)
- The Deputy Mayor (Councillor David McWilliams)
- The Town Clerk (Mr Ernest W Tame0)
- The Chief Constable (Captain A. C. Dawson)
- The Chief Librarian (Mr John Shepherd)
- The Architect (Mr Edwin S. Gray)
- The Contractor (Mr E. B. J. Gould)
- Mr Williamson and Mr F J Dodd representing the workmen engaged on the building of the Library)

The Mayor handed the King an address of welcome from the corporation.

His Majesty handed the mayor a reply to the address.

The band of the H.M. Grenadier Guards were posted at Kings Square along with the Colour Party and escort of the 4th – 5th battalion of the Cheshire Regiment and detachments of the 2nd Cheshire Field Squadron, Royal Engineers and the British Legion.

The Mayor then handed the following address of welcome from the Corporation:

"To the Kings most excellent Majesty
"MAY IT PLEASE YOUR MAJESTY
"We, the Mayor, Aldermen and Burgesses of the County Borough of Birkenhead desire to offer a loyal and hearty welcome to your Majesty and to Her Majesty the Queen on the occasion of your visit to Birkenhead in connection with the opening of the new tunnel and roadway under the River Mersey.

"As Your Majesty is aware, the Corporation of Birkenhead have combined with the City of Liverpool in the promotion and completion of this important undertaking. It is our fervent hope that the new tunnel will not only provide facilities for the national well-being, but also assist in the revival of trade in this town.

"Such prosperity as Birkenhead enjoyed in its past days was due principally to shipbuilding and the industries associated with the Birkenhead docks. The recent depression in these industries has been to us a matter for regret and concerns. We are happy to hear witness to the fortitude with which the attendant difficulties have been faced locally by Your Majesty's subjects. We therefore express the hope that this important date in in the town history may prove the beginning of a period of greater prosperity for Birkenhead.

"We desire also respectfully to record our great satisfaction that Your Majesty has been graciously pleased to signify your willingness to open the new central library, which will provide a valuable addition to the public amenities of the borough.

"We are grateful for the benefits derived from Your Majesty's reign of nearly a quarter of a century and we pray that under almighty God's guidance, our Majesty's may be blessed with good health and may long be spared to guide the destinies of a loyal and contented people."

The King said in reply,
"I thank you, Mr Mayor, for your loyal and dutiful address, and for the cordial reception which you have given to the Queen and myself.

"It is with great pleasure that I recognise the public spirit of Birkenhead Corporation in joining with the city of Liverpool in the construction of the Mersey Tunnel which is today being borough into use and I trust that an ample reward will attend this great venture which is promoting the flow of traffic should assist the return of prosperity.

"You have taken a long view of future traffic needs, and your action in engaging upon this project in difficult times bears witness to your faith in the future of the Merseyside which has afforded in the past so many outstanding illustrations of civic and commerce enterprise.

"I am confident that so long as that spirit of initiative prevails among your citizens no opportunity of advancing the welfare of your community will be neglected. In all such efforts, I wish you Gods speed.

"I am fully conscious of the difficulties and discouragement, which the people of Birkenhead have so bravely faced in the past few years.

"The shipbuilding industry on which the prosperity of your city so largely depends has been one of most severely hit by worldwide trade depression, and it is my earnest hope that the improvement in employment in recent months will continue.

As the cars were about to return to the Royal dais, the town Clerk could be seen rushing across to the side of the square nearest to the market. He returned with an elderly gentleman, Mr S. F. Gillingham, Birkenhead Centenarian, who was in his 102nd year.

Mr Gillngham was assisted towards the royal dais by a friend and the Town Clerk, but their King and Queen did not wait for Mr Gillingham to reach the dais as they left the dais and went to meet him. The crown cheered at this act of thoughtfulness and touched by this single honour accorded to Mr Gillingham by the King and Queen, who was very grateful for their greeting.

As the King and Queen returned to the dais, Mr Gillingham, despite his age, was seen to return to the crowd with a perfect straight back due to his very proud moment.

12:30

As the royal party left the dais, it passed a line of ex-servicemen who stood to attention followed by a group of children who stood up and cheered enthusiastically as the royal car passed. The Royal car proceeded along the three-mile journey along Market Place South, Conway Street, Argyle Street, Grange Road, Whetstone Lane and Borough Road.

12:40

At the new £60,000 Central Library, which was designed by Messers Grey Evans and Crossley, the band of the H.M. Coldstream Guards were positioned along with a detachment of the Chester (Earl of Chester) Yeomanry under Major Williams, who acted as Guard of Honour flanked by members of the Birkenhead School O.T.C. and cadets from the training ship Conway.

To assist in keeping the crown entertained whilst they awaited the Royal party to arrive, the band of the H.M. Coldstream Guards played. As the Royal car approached, the crowd began to cheer and carried on for several minutes during the opening ceremony, such was the local excitement.

In front of the Library, there was a pillar which had been topped by a crown which contained a silver gilt case in book form decorated with the borough coat of arms in enamel. It was inscribed, "To commemorate the opening of the Birkenhead Public Library by His Majesty King George V July 18 1934."

The King pressed a silver switch in the cover of the book, a Union Flag over the door fell away and the doors swung open. It was noticed that the Queen displayed the liveliest interest at this form of ceremony and smiled graciously to the cheering crowd. The switch book was handed to His Majesty as a memento by Alderman G. A. Solly chairman of the Libraries Committee. The King expressed his thanks to Alderman Solly and his regret that he could not stay longer, but he was already behind schedule. The King also remarked that the souvenir switch case was very interesting and he was obliged for the gift. Alderman Solly was also presented with a gold pencil by Mr F. Hamer Crossley of Grey, Evans and Crossley.

This magnificent building which was built in a classical style with bold pillars at its entrance and a symmetrical frontage houses around 65,000 books along with a reference library, reading room and a children's library. It was also said to house a cinema lecture hall with will be equipped with the most modern of equipment. The façade is 156 feet long (47.5 metres), built in Portland Stone and set back 70 feet (21.3 metres) from the existing pavement which gradually rises to the library's main steps. It takes the place of the Carnegie Library, which stood on the corner of Chester Street and

Market Place South (Now Kings Square) which now houses the tunnel entrance. It was originally proposed to build a commercial college and a central school, on the adjoining site. I assume this college was eventually built a mile or so up the road and became Borough Road Technical College, a place which I know well as a former student but this is now a housing estate.

12:45

The Royal car then left and proceeded along Borough Road, Bedford Drive, Bedford Avenue, and Bedford Road to Rick Ferry station.

12:55

At Rock Ferry Station, a detachment of the Royal Naval Volunteer Reserve and members of the training ship Indefatigable were posted.

The Mayor, Town Clerk, and Chief Constable were at the station until the departure of the royal train. Their Mayoress of Birkenhead asked her Majesty to accept a bouquet of roses.

Queensway Tunnel Opening Ceremony with Royal party on Platform (Birkenhead Library)

Queensway Tunnel Opening Ceremony (Liverpool Museum)

Commemorations for Tunnel Opening in Cornforth Street (Birkenhead Library)
Martha Rose is seen putting the finishing touches to the painting in the road.
Her son George is seen holding the paint can.
It is worthy of noting that the painting is a section of the tunnel including the four lanes of traffic can be seen as can the proposed (never carried out) tram section below.

Polished Granite Podium Design

Tunnel Mosaic in Haymarket Entrance (Author)

Plaque Reads
THIS REPLICA OF THE ORIGINAL
MAP WAS CONSTRUCTED JULY 1994
TO COMMEMORATE THE DIAMOND
JUBILEE OF QUEENSWAY TUNNEL

Mid River Section of Finished Tunnel (Birkenhead Library)

Today's Tunnel (Liverpool Dock exit) (Author)
Note new cladding system on walls

George Dock Shaft Cover (Author)

Original Toll Booth (Author)

Liverpool Entrance Commemorative Plaque (Author)
(Seen to the right as you enter the tunnel)
Plaque Reads
IN COMMEMORATION OF
ALDERMAN THE RIGHT HONOURABLE
SIR ARCHIBALD T SALVIDGE P.C. K.B.E.
FIRST CHAIRMAN OF THE MERSEY
TUNNEL JOINT COMMITTEE 1925–1928
BY WHOSE VISION FAITH AND COURAGE THIS
TUNNEL WAS CONCEIVED AND CONSTRUCTED.

Statue of King George V in Haymarket Entrance Statues were re-commemorated for the Diamond Jubilee on July 25th 1994 (Author)

Statue of Queen Mary in Haymarket Entrance Statues were re commemorated for the Diamond Jubilee on July 25th 1994 (Author)

Birkenhead Entrance Somewhat Less Attractive Today through the Marshalling Yard from the 1969 changes. (Author)

King Georges Plaque for Opening Birkenhead Library on July 18th 1934 (Author)
Plaque reads
THIS LIBRARY WAS OPENED TO THE PUBLIC ON THE 18TH JULY 1934
BY HIS MAJESTY KING GEORGE V WHO WAS ACCOMPANIED
BY HER MAJESTY QUEEN MARY

WALLASEY TUNNEL

Chapter 12
Kingsway Tunnel

"My grandfather names the first Mersey Tunnel Queensway in honour of Queen Mary so in honour of my father it is with greatest of pleasure that I declare the second Mersey Tunnel open and name it Kingsway."

OPENING OF SECOND MERSEY TUNNEL
by
HER MAJESTY QUEEN ELIZABETH II 24th June 1971

ORDER OF PROCEEDINGS AT WALLASEY
2:35 p.m.
The Mayor of Wallasey (Alderman H. T. K. Morris), the Mayoress, the Town Clerk of Wallasey (Mr. A.G. Harrison, D.S.C.) and Mrs. Harrison will welcome the Lord Lieutenant of Cheshire (The Rt. Hon. Viscount Leverhulme, T. D., J. P.) and Lady Leverhulme.
2:40 p.m.
Part of the proceedings at Liverpool will be relayed (approx.) over the public-address system to Wallasey.
These proceedings were:
Her Majesty will name the new Tunnel and declare it open.
The opening will be marked by a fanfare of trumpets at Liverpool and the firing of maroons from the Liverpool and Wallasey sides of the River Mersey.
The Lord Bishop of Liverpool will dedicate the Tunnel.

The form of service was as follows:
Our Help is in the name of the Lord, who has made heaven and earth. Amen.
We thank thee, O God, for the knowledge, the skill and the labour that have contributed to the fulfilment of this enterprise and we pray that our common life on Mersey-side, our industries and our commerce, may be benefited by it through Jesus Christ our Lord. Amen.
Eternal God, the source of wisdom and charity, who are present in thy power in every place; Bless, we beseech thee, this work of our hands to its proper use; receive into thy protection all who travel by this way, and be their guide and strength, here and to their journey's end; through Jesus Christ our Lord. Amen.
Unto the King eternal, immortal, invisible, the only wise God, be honour and glory forever. Amen.

The Chairman of the Mersey Tunnel Joint Committee will ask Her Majesty to accept a gift to mark the occasion. The gift will be a silver model of the 'mole' used to excavate the Tunnel.
2:57 p.m.

The relay from Liverpool will end. (approx.)

Her Majesty will leave Liverpool at 3 p.m. and make the first official journey through the new Tunnel.

At 3:05 p.m.

Her Majesty will arrive at the Wallasey Toll Area accompanied by the Chairman of the Mersey Tunnel Joint Committee (Alderman H. Macdonald Steward)

The Queen will be met by the Lord Lieutenant of Cheshire (The Right Honourable the Viscount Leverhulme, T. D., J.P.) who will present to Her Majesty the Mayor of Wallasey (Alderman H. T. K. Morris), the Mayoress, the Town Clerk (Mr. A.G. Harrison, D.S.C.) and other distinguished persons.

The Mayor of Wallasey will present to Her Majesty the Member of Parliament for the Wallasey Constituency (The Right Honourable A. E. Marples, M. P.) and some leading members and chief officials of Wallasey Council.

Her Majesty will mount the dais.

The Band will play the National Anthem.

Miss Shirley Wilson (a Wallasey Schoolgirl) will present Her Majesty with a bouquet.

The Mayor of Wallasey will welcome Her Majesty to Wallasey and invite her to unveil a plaque commemorating her journey through the Tunnel.

Her Majesty will unveil the plaque.

The Chairman of the Mersey Tunnel Joint Committee will thank Her Majesty and invite her to inspect the Administrative Building and Control Room.

Her Majesty will leave the dais and the Chairman of the Mersey Tunnel Joint Committee will present to her the Birkenhead and Wallasey members of the Joint Committee.

Her Majesty will then inspect the Administrative Building and Control Room.

On leaving the Administrative Building the Chairman of the Mersey Tunnel Joint Committee will present to Her Majesty representatives of the Consultant Engineers and Contractors engaged on the construction of the Tunnel, employees engaged on the Tunnel construction, and employees of the Mersey Tunnel Joint Committee.

1. Her Majesty will leave the Wallasey Toll area to return to Liverpool.

Following the speech, the Lord Mayor, Alderman Charles Cowlin presented guests to the Queen along with the employees who were building the tunnel. The Queen also inspected the Honour Guard from the first Battalion Lancastrian Volunteers. The tunnel was dedicated by the Bishop of Liverpool the Right Reverend Stuart Blanch.

Although, the second tube of the Wallasey tunnel was yet to be constructed, the tunnel was officially in use and years of planning and construction of the second river crossing had was now complete. The second tunnel opened on 13 February 1974 with little fanfare and an estimated cost of £8.6 Million (£92,943,321.80).

The tunnel is a twin tube tunnel with two traffic lanes per tube and each tube is 3.65 m in width. Prior to the opening, there was a lot of intrigue as to the name of the new tunnel, which was shrouded in secrecy. Some of the suggestions included Dukes Way, Princes Way, Charles Way, and others which were a little more tongue in cheek Mersey Mile, Link Way, Regal Way, Tunnel 2, Mersey Drive, Export Drive, Liver Drive, and Wallpool Tunnel.

The name of the second tunnel was Liverpool's best kept secret and was finally revealed when Queen Elizabeth pressed a button to release the long blue curtains fell away to reveal the 31-foot-long (9.45 metres) name inscribed into Green Westmoreland

Slate and with gold leaf inlay was KINGSWAY. The opening was a colourful yet simple ceremony watched by 6000 people and this included 200 schoolchildren.

At the official opening on 24 June 1971, Her Majesty, Queen Elizabeth gave the following speech,

"I am delighted to be on Merseyside today, both in Liverpool and very shortly in Wallasey. The deep water of the River Mersey has been the foundation of Merseyside's growth. It has attracted shipping encouraged trade, and helped to give Merseyside its own life and personality. It has been a motorway to the world. But the river has also been the problem of carrying people and goods across it. First, there were ferries. There were later supplemented by the rail tunnel. Then in 1934, my grandparents, King George V and Queen Mary, opened the first road tunnel under the Mersey and the effect of the tunnel was dramatic.

"But since 1934, traffic has grown enormously. More than one third of a million vehicles now pass through the tunnel every week.

"The new tunnel is an essential part of the modernisation of the road network. It will give more room for crossing river traffic and it will make a valuable contribution to the life of both communities on both sides of the Mersey.

"The construction of tunnel of this size is always a notable feat of engineering. This one was not accomplished without conquering severe obstacles. The most modern methods have been used including the use of a mechanical mole in the excavation.

"A second tube is already being built alongside. All this is a great achievement for Merseyside can justly be proud. I warmly congratulate the Mersey Tunnel Committee the architects, engineers, contractors and everyone concerned in this remarkable venture. My grandfather names the first Mersey Tunnel Queensway in honour of Queen Mary. So in honour of my father, it is with greatest of pleasure that I declare the second Mersey Tunnel open and name it Kingsway.

In early 1959, a local authority conference on cross-river transport discussed the options on a new bridge over the River Mersey or a new road tunnel under the river. It was decided that both options would be explored. The authority employed Megaw and Brown to undertake the report along with W. S. Atkins and Partners to undertake a limited traffic survey. The report looked at a six tunnel schemes and two bridge schemes with the favourite option falling for a six-lane bridge scheme. This was to be followed by a two-lane tunnel linking the existing branch tunnels. A tunnel could be built more economical as a two-lane system compared to a double two-lane bridge the cost would be of little difference.

The bridge scheme did not gain too many approvals as it proposed 4,500 feet (1371.6 metres) span with six lanes. Following a number of meetings, a steering committee was set up in early 1962 to look at the options available. The committee was made up with representatives from Cheshire, Lancashire, Liverpool, Birkenhead, Wallasey, and Bootle Councils. The boroughs of Bebington, Crosby, Ellesmere Port, Mersey Tunnel Joint Committee, Mersey Docks and Harbour Board, British Transport Commission, and the Ministry of Transport. The committee may have taken heart that a new bridge on the other end of the river between Runcorn and Widnes had opened in 1961.

The options to be considered by the Steering Committee included the combination with road, tail, ferry and transport services within the Merseyside area. From this a number of sub committees were set up to further investigate detailed traffic surveys, and analysis through the new computer programs. Traffic problems on both options were considered and this would include collecting tolls and clearance for shipping within the river and not affecting what is a very busy shipping port. If the bridge was to be considered, it would have to be of such a height to allow shipping under the bridge at high tide. This would also see a long approach to the tunnel to allow the traffic to gain sufficient height to proceed over the bridge.

In September 1963, the Mersey River Crossing Committee drafted an interim report with the following recommendations. The provision of a tunnelling preference to the bridge options provided greater flexibility in timing, traffic distribution and capital outlay and a capacity for tidal traffic with 12 feet (3.65 metres) wide lanes. There was also the consideration that the bridge proposals would not fit in with the new road plans for the overall area.

If the need arises in the future, a tunnel could be duplicated with the capital outlay being spread over a longer period. This was to prove a major point as the river is a very busy port for all manner of goods and oil and is about to get bigger when Peel Ports expand the Seaforth Container Port in Liverpool in 2014

Urgent preparation of a scheme for a new two-lane tunnel was put into planning. Consideration for using the old disused railway route from Seacombe was made for the possible Wallasey (Wirral) entrance to the tunnel. This railway used to take the public all the way to the Wallasey (Seacombe) ferry terminal only two minutes' walk from train to ferry. It also has Liscard and Poulton Stations within the cutting further up the line from the ferry terminal, close to the Mill Lane Bridge, with a connection with the present day Wirral line around the area of Bidston Station.

Unfortunately, passenger receipts were poor and the line closed in 1960, which resulted in it laying derelict until the Wallasey Tunnel Scheme arrived in 1966. Siting of the tunnel portal in Liverpool would be to the north of the proposed new inner ring road. The recommendations seemed favourable to the proposed scheme which showed the most northerly of the six lane tunnel proposals and a modified Liverpool and a re-planned inner city road system. Parliamentary powers were requested for the construction of a tunnel and it was also recommended that Mott Hay and Anderson were to be the Engineers.

In November 1964, a Bill was lodged in Parliament for the use of the Seacombe rail cutting and the Liverpool portal was as per the plans of 1964.

Although, this was strongly supported locally, the Bill was not unopposed. As with the first tunnel, we now had various councils fighting against each other for their own respective views and those of their constituents. Birkenhead preferred the tunnel linking with the existing branch tunnels, but Wallasey market gardeners opposed this as they objected to part of the proposed road layout.

The summer of 1964 was the start of preparations for the tunnel. This early start was to ensure the work would be carried out in a timely and efficient manner following on from formal Parliamentary approval. This pre-contract and construction work included the preparation of the contract documents and all information and drawings necessary to form those contracts.

In 1965, during the passage of the Parliamentary Bill a master program was prepared in anticipation of Royal Assent that summer followed by authority to proceed in the autumn, with a start on contract works at the beginning of 1966 with the receiving Royal Ascent in 1968. Based on the preliminary designs prepared at the time, a construction period of 4 years 8 months was envisaged, with a completion of the project by autumn 1970, providing that the relevant authorisation was obtained at the appropriate times and that there were no major unforeseen difficulties.

It was realised at an early stage that the acquisition of land, demolition of buildings, rehousing of displaced families and relocation of business would have to be phased beyond the start of work and in fact, these operations took three years to complete. The early programming therefore provided valuable advance information on the priorities of acquisition and made allowance for approach road contracts to start with limited working sites, increasing in scope of works as more sites became available.

However, in June of 1969, Alderman Hugh Platt, leader of Birkenhead Council, and Deputy Chair of the Mersey Tunnel Joint Committee, asked if it was necessary to duplicate the tunnel. Latest figures show that in April of 1969 a total of 1,599,953 vehicles used the existing Mersey Tunnel. This were 62,000 fewer vehicles using the tunnel than the previous year. The Mersey Tunnel Joint Committee, however, was expected to press for immediate duplication of the second tunnel following the publication of the Malts Report in late June of 1969.

Birkenhead council was not expected to oppose the tunnel but Alderman Platt stated that it was a matter of priorities of which should come first, a second tunnel or Merseyrail loop system. He went on to state that,

"In view of the latest tunnel figures duplication of the second Mersey Tunnel is not the right priority. Once the second tunnel is opened it will only add around 20% and this includes new approach roads, on both sides of the river, and a bye pass in Rock Ferry."

Alderman Platt told the committee:

"The usage of the tunnels would continue to decline due to the saturation of parking in Liverpool. People are already seeking alternative means of crossing the river and the railway is the obvious one. At this time the Bill for the loop line had been passed in Parliament and the Chairman of the Mersey Tunnel Joint Committee"

Alderman H Macdonald Stewart, leader of Liverpool City Council stated that:

"The twinning of the tunnels was always envisaged from the outset. I can't see anything in these monthly figures to suggest any difference and it could just be a poor Easter or something like that. I am taking no hard and fast line on this but we will have to see what recommendations Malts makes."

The sequence of construction and times of completion of the approach road contracts were considerable affected by the statutory services (Gas, Water and Electricity) main and cables, which had to be supported and diverted, and by the necessity of maintaining uninterrupted traffic flows across the works. The master programme took account of these restraints and also gave a forecast of phasing of financial requirements during the construction period.

From the master programme, more detailed programmes for each contract were produced to form part of the contract documents, providing an overall control within which each appointed contractor was required to progress the work as it was allocated to the contractor. Critical path analysis techniques were adopted to monitor progress regularly.

At monthly reviews of progress, the programmes were updated, consequently effects on the project as a whole were examined and action was taken when critical work was falling behind. Each contractor was informed of the changing order of priorities as the work progressed, and the predicted completion of the project. In the event, the initial programmed progress on tunnel construction could not be sustained and was irretrievably delayed by labour disputes and by the two major events in the tunnel drive. The actual construction period was 5 years 5 months, with the tunnel opening on 24 June 1971 – less than 6 years since authorisation to the final completion of the tunnel.

The tunnels are circular in cross section with an internal diameter of 9.63 m and a composite reinforced concrete & steel segmental which are painted to provide the final finish for the road tunnel. Welded steel strips cover the lining joints between the segments.

A Robbins Tunnel Boring Machine (TBM) excavated an 11.2m diameter tunnel, the majority of which was bored through the underlying sandstone rock formation. The South Tunnel was the first to be driven. The TBM was then turned around and driven towards Liverpool. The TBM was known as the Mangla Dam mole after its previous work on the dam in Pakistan.

Under the River Mersey, the top or crown of the tunnel varies between 7 m to 15 m. Cut and a cover section approaches at either end of the tunnel were constructed through superficial soils and boulder. The South Tunnel has an oversized section of tunnel at mid-river that forms a 35 m long, 12.37 m wide and 10.76 m high emergency lay-by. This was constructed by enlarging the tunnel locally and lining it with an elliptical shaped cast iron lining. The original configuration of the Kingsway Tunnels incorporated links between the two road tunnels at the ventilation shafts, approximately 460 m from the tunnel portals. Included in this was an additional link below deck level at the mid-river emergency lay-by. The passages located at the ventilation shafts with a single door opening at walkway level with an indirect and narrow route to the adjacent road tunnel.

Unfortunately, this now meant that the existing tunnel contracts would have to be renegotiated, as this would prove to be a cheaper and more efficient proposal than going out to tender for the second two-lane tunnel. It would also alleviate the possibility of having two contractors on the separate tunnels during construction. The new contracts were put on hold until the Merseyside Area Land and Transportation Survey had advanced to such a stage that would allow definitive recommendations to be made. The final report would prove to be in favour of the second tunnel and negotiations with Nuttall Atkinson were agreed in the summer of 1969.

An Engineers Report of March 1965 to the Committee for the potential development stated,

The provision of dual carriageways for a two-lane tunnel is advisable both in permitting free flow of traffic to and from the tunnel and in providing, during rush hour, substantial reservoir capacity without causing congestion in other streets.

The path of the new tunnel was to be planned with the second tunnel next to it so as to enable its use during times of congestion. This would all depend on an adequate approach and the use of the mid Wirral motorway (M53).

The new tunnels would enhance the importance of the industry on both sides or the river and the region as a whole. This was to prove a vital link in the war effort for

World War 2 and the Atlantic Convoys. Vital supplies of men and materials came across the German submarine wolf pack infested Atlantic on a daily basis. This second Mersey Crossing would be designed to keep traffic away from the centre of Liverpool and Birkenhead.

The rock that would be cut through to form the tunnel was known to be reasonably uniform and sufficiently self-supporting so as not to require large amounts of additional support. However, the amount of water that was expected to be present within the tunnel during its construction, through natural seepage from the rock was getting close to the maximum allowable for using a TBM. The use of a TBM would allow a smooth bore for the tunnel rather than a rough bore of the explosives excavation. This second method would have additional works attached to it in that the rough edges would have to be smoothed out to allow a good seal with the tunnel walls. The TBM would alleviate this and allow the tunnel wall to be smoothly adhered to the tunnelled sections.

This development and possible method of construction was to be left open in the tender documents and when the tenders were returned it was obvious the companies that were asked to tender had chosen conflicting options.

However, it was specified that tenders for the tunnel should be for excavation by mechanical means without the use of explosives except as explicitly authorised. The lowest tender was for the normal method of construction and the next lowest had gone for a TBM method of construction. The choice of who was to be the successful tenderer came down to cost and time, and this would prove to be a difficult call.

The type of wall lining was now to be a consideration as cast iron was expensive and difficult to lay, whereas concrete was cheaper and would prove to be an easier option. The decorative finished to the concrete would prove to be cheaper than the false lining required for the cast iron sections. The sections, much like the cast iron, could be manufactured in advance off site if necessary. Concrete would be used for the main section of the tunnel which was within the sandstone and cast iron for the Liverpool end which was a mixture of sandstone and boulder clay.

The Wallasey Approach was to use the existing (disused) Seacombe Railway Cutting, which was clear of housing and businesses so there would be no need to compulsory purchase property and relocate any affected businesses or residents. There were disadvantages to this in that the width of the cutting was less than required by the four-lane approach road. The curves of the railway were considerably less than those of required for motorway traffic and the over bridges were lower and much narrower.

On the Liverpool Approach, the filled in section of the Leeds Liverpool canal as well as the old abandoned Waterloo Dock Goods Yard was available for clearance and demolition. This would allow the exit and approach to miss central Liverpool and proceed along its way to the proposed Inner ring road and out of the city. The exit to the tunnels on the Wirral side allowed for crawler lanes (taking the two to five lanes) to allow fully laden heavy goods vehicles to proceed through the tollbooths and allow faster traffic to overtake, thus allowing free flow of traffic through the tunnels.

A central lay-by was added to the tunnel at its lowest point to allow vehicles to be parked during busy periods and the tunnel to flow freely. The parked vehicle would be moved at a quieter period when tunnel traffic flow allowed one lane to be closed to facilitate the vehicles removal. If the breakdown was enough to close the full tunnel, traffic would be diverted through the other tunnel. This would mean a little congestion as there would now only be one lane each way. An alternative solution would be to divert traffic through the Birkenhead Queensway until the obstruction had been cleared. This has happened on a number of occasions, and shows the flexibility of the construction and allowing traffic to traverse under the River Mersey to its final destination.

The toll area would be on the Wirral side as this would allow the existing practice of collection of tolls to proceed and thus avoid any unnecessary confusion in accounting, and it would also allow the old railway goods yard of the disused railway to be used for the tolls and entrance / exist from the tunnel. This area is 217 feet (66.14 metres) wide and had an existing layout in width and level area for the twelve toll booths.

The alignment, levels and gradients that were proposed for the new tunnel would be determined by the old Seacombe railway cutting along with the riverbed and final point of exit from the tunnel. The tunnel was to be constructed by a The TBM, which would be the largest in the world at 45 feet (13.7 metres) long and weighed approximately 350 tons with 55 cutter heads on the main cutting face to drive through the bunter sandstone and this would be a more efficient method than the hand built tunnel of its predecessor. The tunnel lining was to be pre-cast reinforced concrete with a welded steel inner face. Cast iron segments were to be used where the tunnel rose out of the rock into boulder clay.

The road deck within the tunnel is a reinforced concrete construction with pre-stressed beams. Ventilation shafts and ducts connect the tunnel to the two ventilation stations (one either side of the river). Drainage and sump pumps are positioned within reinforced concrete chambers. An area on the Liverpool side has a 50–60 feet (15.24–18.28 metres) cut and cover to form part of the tunnel. This cut and cover had 100 feet (30.48 metres) span resting on a wall of 8 feet (2.4 metres) bored piles. The decorative finishes are of epoxy spray applied to the main primary lining where welded steel is used. Vitreous enamelled steel clad panels above each walkway.

The Liverpool tunnel emerges completely from the rock along with the cutting and the Wallasey portal in sandstone for the lower half and the retaining wall virtually covers the self-supporting rock. Given the ground conditions, the method of construction through the Triassic Sandstone, boulder clay, peat, soft clays and sand was to be given a considerable amount of thought. In Bidston Moss (Wallasey Approach and M53 connection), construction would have to contend with rock which had been eroded locally to 200 feet (60.96 metres) and was covered in peat, soft clays and sand and the River Fender Valley. The new tunnel would be approximately a mile further up river from the existing Queensway and be constructed in a more direct route across the river.

The new tunnels would have a diameter of 31 feet 7 inches (9.62 metres) as an overall measurement. The two lanes in each tunnel would be 12 feet (3.65 metres) each with a height of 16 feet 6 inches (5.0 metres). The tunnel is a circular type with a small space under the roadway for a fresh air supply. The levels within the tunnel have a gradient of 1:30 (3.33%) on the Liverpool side and 1:25 (4%) on the Wallasey side and have vertical curves of 5000 feet (1,524 metres) radius. The tunnel was to have a rock cover of better than 20 feet (6.09 metres). The opening portal positions of the tunnel are fixed at the lowest practicable road level. Liverpool has a depth of 58 feet (17.67 metres) and Wallasey 71 feet (21.64 metres).

Quote from Michael Chalton) who viewed the tunnel during construction

"I'd been bitten by the underground bug for a few years, going mostly down the slate mines in the Blaenau area and lead mines in the Maeshafn area. I'd read about the preliminary works for the new tunnel and had had a sniff at the shaft site (later to become the Seacombe Vent shaft). One day I just walked in to the

temp. office by the shaft – it was in one of the existing old houses, later to be demolished as the ventilation building was erected, and asked to see 'The Boss'. I met Bob Baird. He was in charge of the land drive (pilot heading) on the Wirral side i.e. Seacombe shaft to mid river. A really nice guy. He found time to talk and explain the work and showed me the rate of progress on charts pinned to the wall. I asked if I and a couple of work colleagues could have a trip down. I explained that most of us had been underground before in disused mines etc. and had helmets, lamps etc. I 'think' I had to write an official letter to the contactors 'to keep things straight'.

"A Saturday morning a few weeks down the line duly arrived and we trogged into Bob's office and donned boiler suits and wellingtons (ours) yellow oilskin jackets (theirs) and our own Oldham Lamps and helmets. After a quick talk about what we would be seeing, we walked to the shaft head. We had to wait a few minutes for the crane to hoist a large kibble up with some men in. No enclosed cage with door interlocks or any such nancy stuff then! The kibble was held steady and we got in. Two trips were needed to get us down I think. Stepped out at shaft bottom. The pilot bore was (at a guess) about 10' dia. The bore went landward but we didn't venture up there. We were led towards the river bed. Funny enough, although a narrow-gauge fan, I can't remember anything about any track or tubs etc. I guess there must have been a railway set-up for mucking out. I didn't see any women or children bent double with baskets of muck on their backs! I digress… I do remember a long string of 110v lights going off into the distance. Visibility wasn't great due to high humidity. All driven through the solid sandstone (Bunter Series???). Quite wet with many spurts of water jetting in or just dribbling down the sidewalls. There was a gully down one side to guide to water to sumps at regular intervals, were it was pumped to shaft bottom. Presumably another lift to surface only I can't recall the discharge point.

"An interesting thing which was pointed out to us was that the water in the sumps was 100% Mersey. Yep, tasted salty BUT it was crystal clear! A slight 'blue lagoon' tinge but as clear as could be. The sandstone was a Grade A filter! I was later to see this in a much bigger sump at the bottom of the Mersey Railway pumping shaft at Shore Road. Amazing! We trogged to just short of the working face.

"I think at that time there was still about 150 yds to go to 'breakthrough' where they would meet the pilot tunnel being driven from the L'pool side. Bob said that if all was quiet on 'his' side, you could faintly hear the drilling from the opposing team! I do remember bringing up the point of that in the quiet non-traffic times in the early hours in the Mersey Railway tunnel, it is said that ships propellers can be heard passing overhead. He'd not heard of that and he said that he'd never heard anything from the pilot tunnel. We trogged back to shaft bottom and we were hauled up in the kibble to grass. We were invited back to the office for a cuppa, a natter and a wash.

"Sad to relate, it was in the L'pool Echo sometime later, that he had been killed/run over by someone out for a jolly in a speedboat nr. Porthmadog. Killed instantly. What a bummer! I have valued memories of a guy – I'd guess he was only in his early 30s. Who took the time and trouble to show round a group of non-tunnellers, non-engineers who he didn't know.

"I've lived in Ayrshire for 30 years now, but head 'home' to the Wirral a couple of times a year. If I use the Wallasey Tunnel, I always remember the day I

went on the trip. For a long time, I kept a small chunk of sandstone I picked up from the tunnel floor as a memento. Sadly, it went missing in a house move."

The concrete precast segments were cast in the quarry at Penrhyn. The company who was overseeing the concrete quality control had two members of staff at the quarry, 1 of which, was a gentleman called John Macready who unfortunately had a very nasty accident when he fell through a board placed over a man hole, badly damaging his back.

The Penrhyn Slate Quarry is a slate quarry located near Bethesda in north Wales. At the end of the nineteenth century it was the world's largest slate quarry; the main pit is nearly 1 mile (1.6 km) long and 1,200 feet (370 metres) deep, and it was worked by nearly 3,000 quarrymen.

The quarry was first developed in the 1770s by Richard Pennant, later Baron Penrhyn, although, it is likely that small-scale slate extraction on the site began considerably earlier. Much of this early working was for domestic use only as no large-scale transport infrastructure was developed until Pennant's involvement. From then on, slates from the quarry were transported to the sea at Port Penrhyn on the narrow gauge Penrhyn Quarry Railway built in 1798, one of the earliest railway lines. In the 19th century, the Penrhyn Quarry, along with the Dinorwic Quarry, dominated the Welsh slate industry.

The quarry holds a significant place in the history of the British Labour Movement as the site of two prolonged strikes by workers demanding better pay and safer conditions. The first strike lasted eleven months in 1896. The second began on 22 November 1900 and lasted for three years. Known as 'The Great Strike of Penrhyn" this was the longest dispute in British industrial history.

In the longer term, the dispute cast the shadow of unreliability on the North Welsh slate industry, causing orders to drop sharply and thousands of workers to be laid off.

From 1964 until 2007, it was owned and operated by Alfred McAlpine plc. In 2007, it was purchased by Kevin Lagan (an Irish businessman who is the owner and chairman of the Lagan Group) and renamed Welsh Slate Ltd.

Kevin Lagan and his son Peter (MD of Lagan Building Solutions Ltd) are now directors of Welsh Slate Ltd, which also includes the Oakeley quarry in Blaenau Ffestiniog, the Cwt Y Bugail quarry.

The tunnel segments were transported to Nutall-Atkinson's compound, on the site of the old swings. They were lowered to the portal area and transported to the TBM by single cab Foden dump tricks. According to those in the tunnel at the time, the tunnel invert was not the safest place to be. The concrete test cubes were taken to the Sangberg and British Inspecting Engineers laboratory in Gorsey Lane, Wallasey for testing which also undertook the testing for the concrete on the Liverpool side and the Wallasey Approach roads.

The project for the second River Mersey Crossing could be seen as a more complex project than the first tunnel. Although, the tunnel is shorter, the links to it on both sides of the river proved to be large engineering feats in their own right. Work was to start in the autumn, based on the preliminary designs a construction period of less than five years which was all down to the project receiving all relevant authorisations and there were no major unforeseen difficulties in planning or construction of the project. Avoidance of noise during the tunnels construction, particularly in built up areas on the Wirral side as they were right next to the approach roads and tunnel route. This would be achieved by using a TBM and it would also allow for a more accurate bore and placement of modular tunnel lining pieces, thus saving time and money on the construction.

From the master plan, more detailed plans for each contract were produced to form part of the contract documents, providing an overall control within which each appointed contractor was required to progress the work allocated. By using the master plan to disseminate the projects requirements, it was possible to now make definitive smaller plans for the tunnels construction, which would be linked into the master plan.

From the smaller plans, it was now possible to get detailed critical path analysis charts together. This form of chart is routinely used in construction and shows the various staged of the project along with the projected start and finish date for each stage. In areas that are critical, these are highlighted and will also show which of the areas (if any) they are linked to. This allows all parties to see if the specified period is to be met or arrangements have to be made to ensure the dates are met.

The contractor would normally call weekly or monthly reviews of progress, and ensure everyone is keeping to the plan and make any necessary arrangements in the supply of men or materials. The effects on the project as a whole would be examined and any action would be taken when critical work was falling behind. In the event, the initial plans for the progress of the tunnel could not be sustained and was irretrievably delayed by labour disputes and by the two major events in the tunnel drive. The actual construction period was 5 years 5 months, with the tunnel opening on 24 June 1971—less than 6 years from authorisation to completion.

In planning the contracts, the companies had to be aware that the Tunnels are 27 m centre to centre and the various contracts for its construction were allocated as follows.

The initial project plan envisaged the allocation of work into eight main contracts.
1. Pilot Tunnel 1965 – Marples Ridgeway Ltd with work starting in January 1966
2. Main Tunnel March 1967 – Nuttall Atkinson and Co alongside Sir Alfred McAlpine and Son Ltd.
3. Liverpool Approach Roads 1967 – Marples Ridgeway Ltd
4. Wallasey Approach March 1967 – Nuttall Atkinson and Co alongside Sir Alfred McAlpine and Son Ltd.

Ventilation Buildings
Administration Buildings
Electrical and Mechanical Plant and Equipment

Wallasey Approach and a 1.25 mile section of the new M53 link road across Bidston Moss – 1967 Sir Alfred McAlpine and Son Ltd along with Leonard Fairclough Ltd.

All bridges, retaining walls and landscaping were given to Mr T S Harker of Bradshaw Rowse and Harker as a Consultant Architect and partner in the company. He was also an advisor on the architectural features within the first (Queensway) tunnel

A tunnel finishes contract was awarded to Nuttall Atkinson & Co in July 1970, which allowed the finishes of the tunnel which would be closely followed behind the road deck construction. This compression or critical path association of work is essential to meet the June 1971 opening date.

To allow a free flow of the works, a series of nominated sub-contractors were appointed and their work was generally carried out in the order in which it is described below:

Sub-contract 9S (Blakely Ltd).

Provision and fixing of preformed cleats, brackets and channels welded to inner mild steel face of primary lining.

Sub-contract 9T (Acalor (1984) Ltd and Brook-Blast Ltd). Grit blasting and application of protective and decorative paint to inner mild steel face of primary lining and welded fixings.

Sub-contract 9P (H J Godwin Ltd). Supply and installation of pumping plant at mid-river, ventilating shafts, portal chambers and pumping stations on each approach.

Sub-contract 9D (UK Construction and Engineering Co Ltd) Provision, laying and fixing of cast iron drainage pipes and fittings in tunnel invert and mild steel pumping mains, and fittings at walkways, pumping stations and shafts.

Sub-contract 9F (Carter Horseley Engineers Ltd). Provision and installation of secondary lining to cast iron tunnel and vertical dado panels throughout tunnel, including provision for fire points, electrical distribution panels and access doors; provision and erection of mild steel ladders and platforms in access shafts and tubular handrails throughout tunnel.

Sub-contract 9C (Tysons Contractors Ltd). Provision and fixing of precast concrete walkway units throughout tunnel (following installation of pumping mains and hydrants under Sub-contracts 9D) and certain in situ concrete finishing works to walkways; construction of in situ concrete arch ribs spanning between retaining walls at Liverpool entrance to support sun visor, and provision and fixing of precast concrete sun visor units.

Contract 5 (Davidson & Co Ltd). Supply manufacture, delivery, erection, etc. of ventilation plant, comprising four main supply fans and associated ancillary equipment.

Contract 6 (Campbell & Isherwood Ltd). The provision and fixing of all electrical services, cabling, etc., throughout the tunnel.

A large-scale project such as the Mersey Tunnel requires a large number of organisations, and specialist contractors. All of this has to be pulled together and one overall person or group in charge. With this project, you have both tunnels, approach roads on both sides of the river, links with the Department of Transport for link roads and central Wirral Motorway (M53).

The tunnel and approach roads were mainly constructed for the Mersey Tunnel Joint Committee but the project also included some principal roads which were constructed for both the Wallasey and Liverpool Corporations, for whom the Mersey Tunnel Joint Committee acted as Agent.

The viaducts and roadworks at Bidston Moss were constructed for the DoE (Department of the Environment) as part of the trunk road plan. This contract was administrated by the Divisional Road Engineer for the North West on behalf of the DoE, who acted as Agent for the Wallasey Corporation in respect of the principal roads and for the Mersey Tunnel Joint Committee in respect on the tunnel approach element.

The Consulting Engineers were acting on behalf of four Clients, whose joint and separate interests were to be taken into account in the supervisory organisation and procedures.

Project control, on such a complex project would normally fall to the Consulting Engineers who would dealt with planning, design, preparation and administration of the contracts, liaison with the Clients and the organisation of the site supervision. This would allow a central point for all information and ensure two separate parties are not going in opposite directions to reach the final goal for the project.

The Architect to the Mersey Tunnel Joint Committee prepared designs and contracts for the surface buildings to meet the day-to-day requirements of the tunnels equipment and day-to-day operations such as vehicle and personnel within the tunnel complex. He was also appointed to assist in the design of the approach structures, the finishing of the tunnel, which would be seen daily by the tunnel users. All of this had to include the requisite planning and building control approvals.

Close liaison between the Consulting Engineers and the Clerk and General Manager of the Mersey Tunnel Joint Committee secured the necessary authorisation and instructions of the Committee as work progressed. Meetings were held at regular intervals with the officers of the local authorities concerned and standard departmental reporting procedures were adopted in communications with the government

As with most major construction projects, the actual scope of the individual contracts has to be seen in respect of the overall project. Given the numerous disciplines and levels of experience, the contracts the timings of the operation had to be maintained with the eventual eleven main contracts.

For the Wallasey Approach contract, it was decided to make approximately two miles of the approach road, part of the main contract. This was to facilitate the contractor being responsible for building the road embankments with material excavated from the tunnel. The contract also included work below existing ground level in the construction of the road deck, pump rooms, portal sumps and ventilation shafts.

A separate contract was given for the Liverpool Approach and this covered all work north of the Liverpool Portal sumps and 800 feet (243.84 metres) of cut-and-cover tunnel section. Also included was the construction of an embankment on the other side of the river, for the future road works on the Bidston Moss. This was to be constructed form the substantial quantities of excavated material within the tunnel.

The first tunnel to be excavated was designated as 2A and was to be excavated. The very hot and difficult conditions encountered within the tunnel were very hard on plant and personnel. This problem also persisted in the construction of tunnel 2B, and shaped the whole course of the works. Perseverance and the application of the contractors experience along with improvements in equipment and techniques brought the machine drive for Tunnel 2A to a successful conclusion and secured a very considerably improved rate of advance in Tunnel 2B.

As with the original tunnel, there was a large element of investigation before any construction work started. In the case of the pilot tunnel, the ground treatment was undertaken to:

- Explore and consolidate weak rock, open fissures and faults;
- Divert water from the crown to the invert of the excavated tunnel profile, thereby improving working conditions in the around the TBM.

The pilot tunnel was completed in January 1967 after some 12 months of work using explosives and hand tools, the pilot tunnel proved the proposed route was viable. One bad fault was found 1,200 feet (365.76 metres) from the Wallasey shore but it was anticipated that this fault would cause no major problems with the tunnels construction.

Nuttall Geotechnical Services Ltd was given the contract to drill out the pilot tunnel, explore the rock and carry out injections over the 4000 feet (1,219 metres) under-river section. This work was finally completed on 1st April 1968 and included the lay-by of tunnel 2B. In the initial stages, Ordinary Portland Cement grout was used, but was later substituted for *Pozament grout*.

The pilot tunnel was to enable the construction of 18 feet (5.48 metres) diameter working shaft sunk on each side of the river. The shafts were to have an offset of 45 feet (13.7 metres) from the centre line of the tunnel. Primarily due to its location, the Wallasey Tunnel was sunk into rock with the upper part in cast iron and the lower unlined in concrete during the construction of the tunnel. The Liverpool Tunnel was to be sunk 16 feet (4.87 metres) into rock and use timber supports and natural rock lining as per the Wallasey Tunnel. Upon completion of the tunnels they were lined with reinforced concrete and used as access and utility chambers.

The pilot tunnel was constructed by explosives with approximately 60 lb (27 kg) of explosives adding around 6 feet (1.82 metres) advancement in the tunnels length. The 7,300 feet (2,225 metres) long pilot tunnel was constructed in wet and arduous conditions in 12 months with an average weekly advance on each face under the river was 64 feet (19.5 metres). The amount of water within the tunnel reached 150,000 gallons per hour as the drive continued. The pilot tunnel contract included for the construction of 7 feet (2.1 metres) diameter mid river sump tunnel to be constructed at the lowest point.

The pilot tunnels were approximately 11 feet 6 inches (3.5 metres) wide and 12 feet (3.65 metres) high and the upper half made into a semi-circular arch. The main supports for this arch were constructed from 6 inch x 5 inch (150 x 125mm) RSJ's (rolled steel joist). Probes were drilled ahead to provide 10 feet (3.04 metres) of cover following excavation. As with the previous Mersey Tunnel, fissure grouting was to be used as the ground conditions were not exact but very similar to those seen in the previous tunnel.

When it came to the planning for tunnel 2B, the decision was taken that a similar pilot tunnel construction should be undertaken for drainage, ventilation and access. It would also prove beneficial as the TBM would have required untested modifications to cut full face without a central pilot tunnel

During construction of the pilot tunnel a two feet (609 mm) wide fault zone was encountered around 4500 feet (1371 metres) into the tunnel. The area was inspected, tightly grouted and supported by timber and steel arches before being re-grouted. The consolidation of the fault would enable the TBM to excavate through by employing temporary support in both the pilot tunnel and the rock crown between the rear of the TBM and the last ring erected

As the TBM approached this area, the contractors removed part of the supports to inspect the area. They found a tacky silt substance along with a cavity approximately 16 feet (4.87 metres) high. Measurements taken at the time showed that this erosion was continuing and that a potentially dangerous condition already existed. The TBM was stopped and a sandbag bulkhead was erected in the pilot tunnel underneath the fault, to provide additional support to the rock and reduce the area through which a possible problem might occur. Concrete was blown into the cavity (as with the original tunnel) and the fault grouted. With the immediate danger overcome, consideration was given to methods of driving the TBM through the fault zone.

There was uncertainty regarding effectiveness of the emergency concreting and grouting was unknown. A scheme to excavate in an upwards direction was used to investigate the stability of the fault and its possible ramifications. This investigation would take a heading over the top of the TBM and investigate until it found sound rock. The results of the investigation would determine the course of action that was to be taken.

Everything being well, the fault would be sufficiently stable and the TBM would be able to proceed as planned. To enable this, the fault would have to be plugged with concrete to the sides and a concrete canopy to support the rack above. This would then allow the TBM to proceed in relatively safety. To provide support to the rock and the advancing TBM through the fault area, a 56 feet (17.0 metres) long concrete collar was concreted in the pilot tunnel. The collar was constructed from corrugated iron sheets which were used as a shutter (frame) and concrete was blown forward from a placer pan located at the rear of the TBM.

Due to the hardness of the rock in certain areas, progress on the tunnel was slow. The main fault was reached on 25th November and was found to be a 42 feet (12.8 metres) from ring 689 the rock was found to be shattered with wide seams filled with silt and sand. As with all tunnel construction, there was water seepage, but this was not too excessive but needed to be stopped.

Inspection of the central heading revealed that the earlier concreting and grouting of the fault void had been successful and a tight seal between the edge of the cavity and the grout was apparent. Also, excavation of the two side headings was practicable. The central heading was reinforced by steel beams placed in the bottom and then concreted by placer and grouted through tubes left in. Excavation of the two side headings started on 6 December, the south heading being advanced first. These were completed by 15 December, including the local plug headings at the fault. Both headings were then reinforced, concreted and grouted. The central access heading over the TBM was then reinforced and concreted back to ring 689 and the key was placed on 31 December, 1968.

Concrete Segments:

Ring Nos	Average Overbreak, feet	Length, feet
1–10	12 x 1.5	40
22–25	12 x 1.5	16
61–63	24 x 1.5	12
66–73	24 x 12	32
74–80	24 x 4.5	28
233-250	12 x 3	72
271–297	12 x 2.5	108
659–689	8 x 2.5	124
763–778	14 x 4	64
820–835	18 x 6	64
946–954	8 x 3.5	36
977–985	20 x 4	36
1185–1188	12 x 3	16
1380–1426	18 x 5	188
1433–1446	18 x 2.5	56
1449–1451	24 x 5	12
1452–1496	24 x 7	180
Total		1084

The TBM, used on the Wallasey Tunnels was a Robbins 371, which had been previously used on the Mangla Dam which was the first of the two dams constructed to strengthen the irrigation system of Pakistan as part of the Indus Basin Project, the other being Tarbela Dam on the River Indus.

The US Company Guy F Atkinson and Co of San Francisco sponsored the project for a consortium of eight American construction companies.

To enable the TBM to be used in the Wallasey Tunnel, the TBM had to be extensively modified and improved during the course of the construction of the tunnels. By the time the Mersey Tunnel were due to start, it has been suggested that all plant from the Mangla Dam had been sold to a plant disposal company in Houston Texas. The TBM had to be repurchased and was shipped in crates and arrived at *Gravesend on 7 July 1966*.

Once the TBM had passed through Customs and transported to the work site, the modifications started. To enable this information be ready for the TBMs arrival and subsequent use in the tunnels, a lot of work in San Francisco and London had to be completed. The redesigned machine was erected and ready for moving forward to the Wallasey Portal face on 18 November 1967.

After the last project in dry conditions in Pakistan, there were detractors who said the TBM would not work in the wet conditions under the Mersey, they were proved to be wrong. Like all pioneers, the engineers and contractors pressed on with their endeavours and their courage and determination prevailed.

The TBM diameter was reduced to 33 feet 11 inches (10.33 metres) with 55 cutters on the cutter head. This was reduced in diameter to 5 feet 6 inches (1.67 metres) to give access to the pilot tunnel. Alterations in the main assembly, as was the propulsion system and the provision of equipment for segment handling and erection. The rear cutting head was equipped to handle the concrete lining segments with a crane and placing boom. By adjusting the machine by a factor of 4 feet (1.2 metres) which was the width of one segment, a maximum of 16 feet (4.8 metres) of tunnel would be unsupported. This would be reduced to 12 feet (3.65 metres) when the concrete segment was in place.

The TBM would move itself through, via propulsion arms which transmit the reaction to a system of thrust blocks, friction gripped on the tunnel lining at axis level, which spreads the load over 20 completed and grouted rings. The TBM was erected between two massive concrete walls, founded on bedrock outside the first portal, and anchored by a heavy horizontal beam extending forward at axis level to the tunnel headwall. The beams were fitted with thrust blocks along their length. This allowed the propulsion arms to the anchor walls.

Following extensive modifications for the wet and muddy conditions the TBM was set to work on a test run on 26th November with it starting to excavate the tunnel on 1 December 1968. Up to 10 February 1968, ring 47 was erected, and the tunnel averaged around 30 feet (9.14 metres) per week. The distance covered was seen as a little disappointing and was partly put down to intermittent trouble with the tunnel crown. It was hoped that once they had past ring 47 the problems would have subsided. However, by ring 53 it was found that the water table within the tunnel had risen approximately 6 feet (1.82 metres) above the tunnels invert. This weakened the strata and a major outbreak of rocks began to rain into the tunnel. These conditions persisted with intermittent outbreaks until 7 March, when better rock was met at ring 83.

Another consequence of the problems was the damage to the TBM cutters. Large lumps of rock could be broken up by the miners in the pilot tunnel but with the TBM, the problems saw damage and even blockages to the waste material chute and conveyors. This saw frequent stoppages and loss of production in the length of tunnel constructed by the day. Steel arch ribs and timber lagging were used to provide ground support in the pilot tunnel and these had to be removed at least 8 feet (2.4 metres) ahead of the TBM. This then allowed shovelling to take place. At the locations of the weak strata and rock falls, the steel ribs and timber arches were trapped and difficult to remove. The additional work to remove the ribs and lagging was not only hazardous but time consuming.

The sheer weight of the broken rock, which is now sitting on the cutter head, caused concern within the tunnel management. One main concern was the possible damage to the main bearing seals and the cutter head. On several occasions, the cutter head became trapped by the rock falls and due to the confined space, access to the cutters for maintenance and inspection was not very practical.

Rock grouting of the under-river section appeared effective at first, but the effect soon declined until it was negligible and the water seepage was not declining. Water was running out of the exposed rock around the TBM and being trapped between the cutter head and the vertical face when the TBM stopped digging. Upon restarting, the cutter head buckets scooped up the combination of water and muck and dumped it onto the conveyor system, which resulted in it cascaded over the machinery. This was causing concern to the contractor and this amount of water and silty sand could have a detrimental effect on the machinery.

As more rock was mined, it caused the conveyors to block at a chute transfer point, with a resulting stoppage of the TBM. More water would then collect at the face and the restarting problem would start all over again. A method of securing this problem had to be found as it would not only have a detrimental effect on the machinery but the whole tunnel programme.

For the first alternative, several methods were attempted:
- Installing a suction pipe in the support between the front vertical steer shoes to pump away the water. The design of the support rendered this impractical.
- Using a 3-inch (75mm) Univac pump with a suction hose threaded down between the rock face and the cutter head. This was unsuccessful due to the difficulty or threading the suction hose past the cutters.
- Grouting through the floor of the pilot tunnel ahead of the TBM. This was very difficult to carry out, due to the flow of water in the pilot tunnel and the depth of silt, which accumulated on its floor, and produced no noticeable improvement in the water situation.
- The use of a drying agent or coagulant deposited as an additive between the cutter head and the rock face a few minutes before digging with reversal of the cutter head for some minutes under full thrust to mix the agent, arising and water, and thus dry or coagulate the slurry. Digging was started immediately after mixing to prevent further ingress of water.

The agents tried were:
Sawdust in varying quantities; both coarse and fine bentonite, using about 400 lb at each start up; a mixture of sodium carboxymethyl cellulose and cement fondue.

Care had to be taken to ensure that the first-floor buckets did not overload the conveyors. It was costly and messy, and the rocks in the coagulated material often blocked the transfer so that the operation had to be restarted.

No satisfactory method was devised for dealing with the water once it had got into the system. Consideration was complicated by:

- The varying quantities of water encountered;

- The need to cope with large rocks, liquid slurry and muck;
- The restricted space available on the TBM;
- The rate at which the waste material is produced;
- The problem if disposing of the silty water.

The TBM chute system was remodelled, to incorporate a scraper conveyor and a sludge box with conveyor. A valve control system was installed directly behind the cutter head support. A valve control system was installed directly behind the cutter head support, and a wide, slower speed 45-degree TBM conveyor. The work-car conveyor was also redesigned and coupled with a pinpoint to the TBM conveyor.

Due to the problems within the tunnel, the grouting of the rings at the tunnel roof was supported temporarily by timber and sand and cement. Back grouting of the tunnel segments was carried out as soon as possible with the ground improvements to provide a tight seal around the ring. Following on from this, the contractor encountered problems erecting the top segments. The contractor decided to alleviate this problem by use of the following.

Bankers Bars – These are welded to the TBM and would span onto the edge of the last ring placed into position. Unfortunately, this system was only partially successful and had a tendency to be torn distort or be torn off during shoving.

Needle Beams – These would be temporary rock props and top segments during erection of the rings. They were drawn out of mountings on the underside of the TBM canopy.

Timber and Steel Packing – This was installed within the cavity over the top of the last ring that had been erected. It gave rock support and cantilevered headboards at intervals to support rock which became exposed during the following shove.

Despite the precautions, the erection of the upper tunnel lining rings could only proceed with careful cleaning one segment at a time. This would all ensure that loose rock is carefully taken care off but progress was slow. The invert in this area consisted of broken rock with beds of silt and sand into which there was a considerable water inflow. The weight of loose rock resting on top of the TBM caused it to settle three inches (75mm) into the tunnel floor.

The contractor tried a number of times to jack up the machine but this proved almost impossible and it was decided to continue the excavation and bring the machine up gradually. This had an effect of the crown of the tunnel which was three inches lower and additional difficulties with the installation of the concrete ring lining. There was also a problem in the grouting of the rings due to a slight distortion of the affected area.

Around rings 53–83, which were located within a residential area 60 feet (18.28 metres) below ground level with 18 feet (5.48 metres) of rock cover and it was over a limited length the over break *(a caving in or loosening of material along the edge of the excavation)* extended up to the base of the overlying boulder clay, but no discernible change in ground levels occurred, and there was no damage to surface property.

Given the current situation, it was decided to pin the rock from the pilot tunnel in advance of the TBM. Initial trials proved that a lattice-work arrangement of inclined hollow bamboo rods was placed and grouted in purpose drilled holes. The bamboo rods were placed in such a fashion so as to be cut off by the TBM cutting teeth as it passed through the tunnel. On 2nd June 1968, the TBM reached the Seacombe Ventilation area (ring 382) and averaged 99 feet (30.17 metres) per week over an 11-week period. The best performing period was the week beginning 26 May when 136 feet (41.45 metres) was covered.

Other points of interest during the construction are the week ending 23 June 1968 208 feet (63.39 metres) was covered at the start of the under-river drive, but the conditions generally deteriorated as the advance proceeded into the river rock and slurry. A period of 15 weeks from 24 June to 06 October *1968*, showed an average run of 52 feet (15.84 metres) per week and 170 rings placed into the tunnel lining.

The following week (week ending 13 October) examination of rock conditions in the pilot tunnel revealed serious deterioration of the rock at a major fault zone. Progress continued with the TBM until 7 November 1968, when the drive was stopped after ring 689. The fault zone continued to be a problem until New Year's Eve 1968 and the TBM traversed the fault zone successfully.

Steady progress was being made and 9th February 1969, seventeen rings were erected, which was best weekly production since August 1968. However, during that week damage to the main cutter head bearing became apparent and on 20th February 1969, when emergency measures had to be put in place to enable works to replace the bearing. This work, which caused a 6 week delay and they had only reached ring 776.

Tunnelling restarted on 2nd April and progress was slow due to constant checking of the tunnel bearing and associated gear box. By 17th April, confidence and moral in the tunnel had been restored and the tunnel workers were up to 17 rings a week. By 27th April 1969, the first permanent cast iron ring was placed in the lay-by of tunnel 2B.

The lay-by is 112 feet (34.13 metres) long tunnel enlargement to provide a 9 feet (2.74 metres) widening over a length of 95 feet (28.95 metres), and 56 cast iron rings. A temporary steel bridge was erected through the lay-by to allow vehicle access to the TBM. An upper deck was erected at axis level for the top work, on which a Coles Hydra 6 ton crane was placed. Two runway beams were positioned at the break-ups at rings 820, 821 and 863, 864, allowing four-face working area.

The rock was excavated by controlled blasting and loaded into skips either by hand or using an Elimco 622 rocker shovel. Skips were hoisted by an air hoist on the runway beams and the rubble was transferred to the TBM conveyor. The Hydra crane was equipped with a hydraulically controlled erector head, to erect and remove segments in addition to its normal use as a crane. Each pair of rings was grouted immediately after erection, and lead caulked later.

Maximum progress was four rings completed in one week of double shifts of 12 hours, eight men per shift. Construction of the lay-by enlargement was completed in May 1970. This was a most difficult and costly construction. Since it lies at the lowest point in the tunnel, the invert was often flooded, to which a mid-river pump was constructed. The men worked in continually wet conditions, in which retaining labour was extremely difficult. The men succumbed to a First World War problem of trench foot. This is caused when feet are constantly sodden and the skin crinkles (not dissimilar to when you have been in the bath to long) and rips easily causing severe injuries.

Steady progress was made, averaging 68 feet per week (20.7 metres), despite the difficult water conditions that existed within the tunnel and on 17th September 1969, grinding noises were heard from the cutter head. Investigation revealed serious mechanical difficulties with the bull pinion tooth. The TBM main drive bull pinions broke away and became lodged between two of the teeth on the ring gear resulting in the following damage occurring;

- All the drive bull pinions were smashed
- The front bulkhead of the TBM cutter head support, in the gearboxes were badly distorted
- The gears and bearings of all gearboxes were extensively damaged;

The bearing adjacent to the bull pinion in each gearbox was driven into the housing formed in the gearbox casing

One ring gear tooth was cracked along the root for two-thirds of its length (the forging of the ring gear is an integral part of the main bearing).

In order to get the works completed on time and enable the TBM to be up and running again, the following works had to be undertaken:
- A new 8½ in (212.5 mm) steel plate and forming the new bull pinions from it (this involved cutting and machining the blanks, gear cutting, and heat treatment);
- Machining the outside of the gear boxes to fit the new registers, as well as forming the new registers out of plate and then welding them onto the straightened front bulkhead to an accuracy of +0.002 inches (0.0508 of a mm. This accuracy was necessary to ensure that the bull pinions meshed correctly with the ring gear).
- Rebuilding the gearboxes, using new parts as required;
- Forming a new seating for the final bearing by machining and sleeving the inside of the gearbox casing;
- Pegging the cracked tooth, using three 3/8 inch (9mm) diameter high tensile socket-headed setscrews.

It took just under three weeks to finally complete the works, and with the first run of the machine, it was realised that a slight miss-alignment of the ring gear teeth had occurred.

After two weeks running, it was noted that there was severe wear on the ring gear and this was found to be a distortion to the bull pinions. After the pinions, had been re-machined and refitted the main bearing and gearboxes were run for the remainder of the drive without further trouble. Progress averaged about 67 feet/week (20.42 m), reaching the TBM extraction shaft area on 4th March 1970, shortly after traversing the complex area of ventilator ducts and service shafts. After starting the tunnel process on 1st December 1967, the 27 months and 5900 feet (1798 metres) of tunnelling process finally holed through on 4th March 1970.

However, on 16th November 1968, during a stoppage at a fault, gearboxes 4 and 6 were removed for inspection, and emulsified oil indicating intrusion of water was observed. The two gearboxes were cleaned and filled with new grease, and the automatic greasing cycle was increased and run continuously. The initial failure appears to have been mainly due to the salt water, sand and silt penetrating into the bearing cavity. Material from the fault was identified in the bearing grease, suggesting that the long delay at the fault position may have played a big part. Steel which could have been a gearbox cover plate, was discovered in the bearing but was judged to have entered between the bull pinions and ring gear when these were already slack, so that it could not have been a primary cause of the breakdown.

Trouble developed again in February 1969 when contamination of the grease was noticed. The bearing was cleaned and degreased, but heavy wear was noted by the contractor. Two days later the bearing was inspected by the Contractor and slithers of metal were discovered, with subsequent analysis indicating that they might have come from a roller cage. On the same day, precautions were taken to reduce stresses and further wear on the bearing, which was refilled with grease, and the TBM was back in service by midnight.

The following day the new grease was inspected and hardened sand was found on the edge of the ring gear. The bearing was again cleaned, recharged with grease and repacked to relieve stress over the next few days.

A total failure of the bearing occurred on the afternoon of 20th February, 1969, when the TBM had advanced 3130 feet (954 metres) and was approximately mid river. It would have proved impractical to remove the TBM from the tunnel and change the bearing. A method had to be devised to ensure this work was completed in the tunnel and completed in complete safety for all concerned.

In the past, similar but smaller bearings had occasionally failed in tunnels, but replacements had been performed outside the tunnel because of the extremely accurate control and almost clinical cleanliness required. A thorough inspection of the surrounding area within the tunnel showed that the rock was sound, so the emergency plans were put into effect immediately and with confidence.

A crated bearing had been supplied when the TBM was shipped was found to be heavily filmed with rust when unpacked, because the protective grease had broken down the dried out. After cleaning, both rollers and races showed pitting and opinions on the lift of this bearing varied from a few to several thousand hours. There was no time for a full repair, so it was decided to source replacements for 40 of the most pitted. The spare bearing was then fitted with a lifting frame and moved into position in a working chamber excavated above the TBM cutter head.

In order to fix the cutter head the cutter head support was retracted with a reverse thrust estimated at 250 tons and 24 x 9 inch (600 x 225mm) Steel Beams were welded to its face and concreted into rock pockets, with 10 x 10in. x 49 lb/feet (250 x 250 x 22kg) Steel columns as rakers (*usually used to prop up a wall and consist of one or more timbers sloping between the face of the structure to be supported and the ground. The most effective support is given if the raker meets the wall at an angle of 60 to 70 degrees. A wall plate is typically used to increase the area of support*).

Further raking struts from the cutter head passed through inspection holes in the cutter head support to the rock behind.

Whilst changing the bearings, two gearboxes and bull pinions were used for accurate positioning and the housing was heated to 11 Degrees Centigrade above ambient temperature. Four 1 inch (25mm) diameter stud bars were used to line the bearing into its housing. New seals were designed, fabricated and fitted, and band saw steel was used to refurbish worn seal tracks followed by extra grease pumps were also added to the existing system.

The TBM's steering controls and tapered dowels protruding from the bearing boltholes were used as guides for the last 4 inches (100mm) of movement. Great care was needed because the lip seals faced the wrong way for the operation. The whole operation was completed in 40 days, against an estimated 45 days.

The progress of the main TBM drive was seriously affected by the rock and water conditions it encountered, and this situation was aggravated by the work necessary to traverse the fault area safely. With the failure of the TBM main bearing, and the discovery of corrosion in the spare bearing, immediate action was necessary to ensure progress in any contingency.

The steel faced concrete segments were designed for machine erection and any change from machine to hand driving presented numerous problems. A scheme was devised to excavate and build on one face by means of a gantry from a new shaft to be sunk close to but landward of the Liverpool ventilation station. This 36 x 15 x 90 feet (10.97 x 4.57 x 27.4 m) deep shaft to be used later for the extraction of the TBM. The top 20 feet (6.09 metres) of shaft was in clay and supported by braced piles with two steel frames. The remainder of the shaft was rock bolted as required, and belled out at the bottom to assist both the settings up of the hand drive gantry and the extraction of the TBM. Shaft construction began on 24 March 1969, and was finally completed in June.

The shaft was a full-scale experiment to ensure the work under the river could be carried forward successfully with the steel faced concrete segments. This would also ensure that in the event of a total breakdown of the TBM, the tunnel could still be worked upon. The following five months were taken up with planning and equipment mobilisation, followed by full-scale work was started at the end of August 1969, driving landward towards the completed drive from the Liverpool Portal.

Progress was soon as predicted, even with the problems associated with hand driving using explosives. Extra care had to be taken near any sensitive areas such as sewers, and possibly causing difficulties with the tunnel grouting. The gantry, gave access to the face for drilling, trimming or support. The three bottom lining segments were placed by hydraulic crane as soon as it cleared, the gantry running on the top of the haunches. Erection of the upper lining segments was by a segment erector at the rear of the gantry like that on the TBM and with a similar curved running beam.

On 29th January 1970, a large rock fall occurred without warning leaving a conical shaped void approximately 27 feet (8.2 metres) in diameter at the base and 25 feet (7.62 metres) high, and 8 feet (2.4 metres) in diameter at the top. The gantry canopy was crushed and the gantry damaged, but no one was hurt. Inspection of the rock suggested that had the TBM been excavating the rock would have fell into the TBM's main areas with not the best consequences and retrieval might have been difficult. The debris was cleared and seating for ribs were cut in the rock throughout the area. Ribs were erected at two feet (609mm) centres on the seating, concreted into the shoulders and close boarded over. Concrete to a depth of five feet (1.52 metres) was placed above the ribs and laggings, followed by further concrete and final grouting from both above and below. Throughout the drive to complete the tunnel, the ribbing above the concrete lining was continued and with luck and good engineering there was no further rock falls. The drive was completed at the end of July 1970, following the repositioning of the TBM at the end of the run a halt from 10 March to 30 April.

The Liverpool Portal cofferdam was completed to its full depth in December 1967 and this allowed the placing of concrete for portal structures to start in January 1968. Work on the tunnel started in February and continued till the end of 1968. This continued on the premise that a high priority would be given to the stage for the main tunnel drive. By the end of April 1968, the special tunnelling gantry, which was really a skeleton shield (known affectionately at the flying bedstead), had been delivered with its ancillary equipment, and erection of the forepart in the portal shaft had begun. When the completed forepart had entered the eye of the tunnel the after part was erected in the shaft. The whole gantry could then move off as one unit; although, the face had to be advanced over 45 feet (13.7 metres) before the gantry tail left clearance for normal muck disposal and normal operation of the system. This stage was reached in mid-June.

The gantry platform travelled along the new tunnel on rollers which were fixed to the cast iron lining as axis level. It was anchored to the lining via hydraulically operated sockets and propelled forward by rams. At the forward end, it carried gun struts to support face timbering or decking, for hand excavation or drilling of rock for light blasting. It also had a roller type erector to place the lining over the upper half of the tunnel.

Excavation over the face was carried out by hand in boulder clay or sand and by light charges of explosive in the rock. A Hymac hydraulic excavator was used to trim the lower part of the face and to excavate and load away the waste material to Edbro lorries. All waste was hoisted up the portal shaft in the Edbro skips and tipped at the shaft head for distribution into the spoil heaps above ground. In loose ground, excavation was completed one ring section at a time, and as better ground was reached space for two rings could be excavated at a time. Grouting equipment was carried on the gantry, and the rings were grouted and caulked shortly after erection.

As the drive advanced, the base level of the boulder clay rose steadily relative to the tunnel, descended in its planned track for the tunnel route. This also increased the proportion of rock and sand to be excavated which proved to be a preferred material to tunnel through. Over much of the drive, the upper part of the sandstone was almost uncemented and this area of the face had to be timbered off the gantry. At 37 feet (11.27 metres) from the portal, the excavation encountered a pocket of creosote at one of the shoulders, requiring precautions against seepage, including attention to the caulking of the rings when built.

Originally, the drive was to have been 600 feet (182.88 metres) long, but due to the poor quality of the upper layers of the rock over the crown and the proximity of the intended terminal point to a railway viaduct above the tunnel, the drive was extended to 670 feet (204.2 metres). Driving was completed in November 1968, and the face was boxed up.

Unlike the Queensway Tunnel, the Kingsway Tunnel lining was designed so that it provides an almost finished tunnel. The concrete surface of the purpose built lining rings gives a substantial saving in cost and along with the welded steel membrane backing allows for a single high standard of decorative finished to be applied. The main road deck construction started on 10 May 1970, and completed on 7 February 1971.

The use of the steel sections was developed by the Welding Institute of Cambridge, who ascertained the difficulties that were foreseen such as water seepage before and after tack welding. During the construction phase, rust on the steel plates affected the adhesive properties of the weld, along with misalignment of tunnel segments. The trials at Cambridge were followed by site trials on segments faced with mild steel and stainless steel plate. These were erected in a particularly wet location in the pilot tunnel to give a true reading of for the tests.

The trials adopted the use of a mild steel plate and gas shielded electrodes. Second was a bare wire gas shield process using 80% argon 20% CO_2 with 3/64 inch (1.17 mm) triple deoxidised wire. Where conditions allowed, pure CO_2 was substituted for the Argon/CO_2 mixture. Standard production segments were tested for external water pressure to 80lb/sq inch within the confines of a specially prepared rig. The concrete sections which would be painted with a black bitumen coating and have caulking grooves and grouting holes have a radial diameter of 31 feet 7 inches (9.62 metres) with a depth of 1 foot (304.8mm) and strength of 5,000 lb/sq inch. The steel lining is ¼ inch (6.25mm) mild steel plate secured to the concrete segments via welding anchors at 9 inch (228mm) centres. To retain the watertight integrity of the steel a cover strip was welded between the segments. The steel is used only in sections above the roadway and not in the lower sections.

As with any tunnel construction, you will have a left and right side concrete section the contractor devised a new system of longitudinal bolts that give continuity and structural rigidity to allow the rings to be cantilevered whilst grouting is being undertaken. Within this new system, there are six types of segment notwithstanding the left and right options which give a total of 14 options

Base Segment A – weighs 3.95 tons and is flattened across the invert to provide 8 feet (2.43 metres) wide plane surface, directly usable by construction vehicles

Haunch Segment B1, B2 – weigh 4.03 tons and are cast with integral launch to carry the road deck, with a ventilation slot which was initially 12 inch square in section but modified to 12 inch (304.8 mm) diameter circle for ease of casting. As with the Queensway Tunnel, air supply is balanced along the length of tunnel by blanking off a proportion of this area.

Segments C, D, E – In unrolled rings two full segments (C) with a circumferential length of 10 feet (3.048 metres) and a similar top segment (D) rest on each of the haunch segments. These are all lined with steel plate. The C and D segments weigh 2.06 and 2.09 tons respectively. In rolled rings half-length segments E weighing 1.40 tons are used on both haunches and to provide break joints followed by C and D segments.

Key Segments – has a circumferential length of 4 feet (1.219 metres) and weighs 1.10 tons. In place of the Concavo-convex joint used elsewhere the joint faces of the key are plane tapering to 1 ¼ inch (31.25 mm) in the 12 inches (304.8 mm) thickness but the D segment faces are slightly convex.

For the section of the Liverpool portal, 680 feet (207.2 metres) of cast iron segments were used.

Special Precast Concrete Tunnel Lining

A large area was needed to manufacture the steel-faced concrete tunnel lining segments. The Tunnel Cement Co made available to the Contractor a disused former quarry site with a level area of about 5 acres and a secondary Quarry for material manufactured and stored during the TBM's bearing change period. This was to ensure constant construction of the segments. The segments were to be held together with through bolts and couplers allowing slight post-tensioning. Development of the lining was closely integrated with the design of the TBM segment erection machinery and the thrust arrangements.

The casting yard was laid out so that all materials (liner plates, reinforcements, cement, sand, ballast, insert tubes, Cofretol tubes, clips, etc.) were fed straight in, processed and passed to the casting lines. Concrete was produced at a central Nuttall-Benford batching plant and Winget Flemert S750 2/3 cu yd pan concrete mixer set-up. It was transported to the beds by agitator lorries fitted with conveyors which reached over the centres of the mould, and vibrated via a poker vibrator. Once the cast segment was finished and sufficiently cured, they were craned to the back of the site for later transportation to the tunnel.

The segments were to conform to tolerances of + 1/16 in (1.56mm) to – 1/3 in (8.3mm) on overall size, and + 1/16 in (1.56mm) to 0 on other dimensions. Great care was required to achieve the desired results. All segments, except the ones which included a haunch, were cast on their sides in pairs. Enough of the casting moulds were provided to cast 14 rings of segments. Heating was installed for cold weather concreting and steam curing was provided.

It was originally envisaged that the regular output of the casting yard would be over 40 rings per week and the delivery of mould parts. In the event, mould preparation was slowed by the need setting the 48 inserts required. The liner-plate hooks and reinforcement was to serve various purposes. To overcome the combined effect and prevent delay to the tunnel construction, two-shifts were formed and work began in February 1968 and continued until November, when 80% of the casting was complete.

With heavy inflows of water from the rock behind the lining, grouting was still susceptible to being diluted and in some cases washed out. An experiment was tried using pea gravel to fill the annular space and then grouting this with a cement grout. It was found exceedingly difficult to make this section of tunnel watertight and it is suspected that the pea gravel was not sufficiently uniform in size and the experiment was abandoned.

The road deck construction was based upon the use of standard inverted T-section precast pre-stressed concrete beams 24 feet (7.31 metres) long, 1 foot 3 inches (381 mm) deep and 1 foot 8inches (508 mm) wide. Mild steel reinforcement 40 mm in diameter was threaded through the steel beams and concrete was poured for the full width of the steel beams. Reinforcing mesh was then laid over the tops of the beams and 4 inches (100 mm) of concrete was then placed over the beams. The concrete was sulphate resisting cement with the aggregates were granite from Pen-Maen-Mawr quarries so Wales had a large part in the construction of the second river crossing.

To enable a good repairable and what may be said workable concrete road deck, the road was divided into 80 feet bays (24.38 metres) which were separated by contraction joints made with a polypropylene former. This form of construction square section style can be seen in many large areas that have used a similar construction. One of the many problems that the contractor had to overcome was that typical motorway equipment used to construct road to normal road standards required by central government. It is easy to use and store the equipment when building a major road network, such as the ends of the tunnels and the link roads, but the tunnel has a very defined and limited space allowance.

Given the complexity and tight spaces within the tunnels, a method of constructing the new roadway, and allow other trades to complete their works on time. All this was to be completed without using the newly concreted roadway. The decision was to erect a Bailey Bridge at the Wallasey Portal to give direct access to the road deck while traffic passed beneath. A Bailey Bridge, which in essence, is a type of portable, pre-fabricated, truss bridge. It was developed by the British during the Second World War for military use and saw extensive use by both British and the American military engineering units. Bailey Bridges continue to be extensively used in civil engineering construction projects and to provide temporary crossings for foot and vehicle traffic.

The road beams were transported by a tractor and trailer running beneath the road deck to the point where they were to be placed. They were handled by a purpose built electrically powered gantry which was fitted with direct drives to four of the wheels. It had a 5-ton self-travelling hoist which worked off a 50 feet (15.24 metres) long beam. This enabled the beams to be handled quickly and efficiently.

To transport the concrete from the mixing Lorries past the uncured (green) bays to the pouring point a system of trolley-mounted conveyors was used. These were from Wickham Engineering and were 12 inches (304.8 mm) wide and ran on an ancillary rail placed outside the bays.

In all, 480 feet (146.3 metres) of this type of conveyor were made available during the tunnels construction. At the discharge point, there was a cross conveyor, spanning the road. This was fitted with a travelling plough so that concrete could be placed across the width of the tunnel in an even layer. The cross conveyor and the adjacent longitudinal conveyor were fixed together and both were powered so that the whole of the discharge end of the system could move along the tunnel with the longitudinal conveyor telescoping under the adjacent system. All of this was closely monitored and controlled via a control panel at the discharge end.

The concrete would then be vibrated (to ensure no air pockets were present in the concrete) by 24 feet (7.3 metres) SGME vibro-finisher. It was then wire brushed to leave a slightly rough finish. This is commonly allowed for to aid traction and adhesion on the road surface by the vehicles using the finished product. The Department of Transport Specification for Road & Bridge Works defines finishes from F1 to F5. (The latter specified as being free from any blemishes & imperfections such as discolouration).

These categories often give rise to considerable disputes between contractors and engineers on site as to what is an 'acceptable' finish. One major achievement of the contractor and the methods used within the tunnel allowed 400 feet (121.9 metres) of deck was cast in 80 feet (24.38 metres) bays in one week. Given the constraints placed upon the contractor and the workforce, this is a considerable achievement for such a short space of time.

The Contract called for construction of an experimental section prior to the main work. This experiment section was completed in the cast iron lined section of tunnel at the Liverpool end between 1 May and 30 June 1969, much valuable experience was gained. The pedestrian walkway was constructed in segments and delivered to site and installed once the roadway was complete. These segments included the ducting and other sections to allow for services to be directly installed.

Drainage of any form of construction is an important factor to be considered, but more so with a tunnel. The last thing the tunnel contractor or owner, be it central government or private company, is water ingress into the tunnel and flooding at the lowest point. With the Kingsway Tunnel, the Liverpool approach disposal of all surface water into the city drainage system was permitted. Within the areas where the city road network joined the tunnel approaches it was straightforward. In the deep cuttings, the surface water has to be pumped out to existing sewers from two main collecting points, at both Limekiln Lane and the Tunnel Portal.

At Limekiln Lane, each of the three pumps had a 1450 gal/min capacity; two were employed as duty pumps and one as a standby, and an ability to cope with a one-year storm. For greater storms, an 18 inch diameter overflow pipe leads from Limekiln Lane sump to the portal chamber, where a storage sump for 125,000 gal is served by three 1100 gal/min capacity pumps operated on the same 2:1 basis.

The storage capacity available at the portal chamber and in the overflow pipeline is sufficient to cope with a storm in the region of 40-year intensity. If this is exceeded the portal sump overflows into the invert of the tunnel and, when the mid-river sump is full it is then stored in the invert at mid-river below the road deck. Precautions have been taken to prevent the flooding of the pump room.

On the Wirral side of the river at Leasowe Road and Bayswater Road, connection of surface water drainage into the existing system was not acceptable. At Bidston Moss, the drainage passes to existing watercourses. For the section between the New Brighton railway and the toll area, two new outfalls had to be constructed into the Wallasey docks. A pump house was required at Gorsey Lane to deal with a low section of the approach, 1200 feet (365.76 metres) long.

For the deep cutting leading to the tunnel, the arrangements at the Wallasey Portal for dealing with storm water are similar to those at Liverpool Portal, each pump has a 1500 gal/min capacity, discharge through pipes which pass down the tunnel, rise to the surface at the Wallasey ventilation building, and end in a new outfall into the River Mersey.

Seepage of ground water through the lining of the tunnel is minimal and external rainwater intercepted at the portals so pumping of water within the tunnel will only come from rain carried in by vehicles. However, as with any major projects day to day management, contingencies for the use of fire hoses, a burst water main, or bypassing of the portal sumps due to blockage or overflow must be provided for. The portal sumps collect water from a very wide area of approach road and, because of potential consequences to the tunnel; provision is made for a 100-year storm. The arrangements relate to the complete system after duplication of the tunnel.

The road deck in the tunnel drains to a mid-river pump of 1440 cubic feet capacity through trapped road gullies connected to a 6-inch (150 mm) diameter drain. This sump is served by two duty and one standby 8 inch (200 mm) vertical centrifugal pumps each with a capacity of 500 gal/min discharging through an 8 in (200 mm) main run beneath the tunnel roadway and up a shaft to discharge into the river at Wallasey. The pump levels are well above the flooding levels of the 100-year storm. Special detector devices and automatic fire precautions are installed to protect the system spillage from tanker vehicles.

Beneath the tunnel, inverts sumps and pumping chambers are provided to handle the large run-off from the wide and lengthy approaches. The portal structures also incorporate vertical shafts up which services (gas, telephones, etc.) will be carried to the surface. In addition to the portal sumps there is a main tunnel sump at the lowest point at mid-river, which comprises 7 feet (2.13 metres) diameter tunnel built at right angles to and beneath and main tunnel.

Above it is a pump chamber formed as a 21 feet 2½ inch (6.46 metres) diameter tunnel accommodating the necessary pumping machinery. When the duplicate tunnel was built, it connected into the same sump and pump chamber, and access between tunnels is available at that point.

By 1960, peak-hour demand was exceeding capacity in the Queensway Tunnel and as the entrances lead directly onto city streets, sever congestion was a serious consideration and growing problem. The main traffic flows from the countryside of Wirral to Liverpool in the morning peak and vice versa in the evening. With only four 9 ft (2.74 metres) wide lanes that can see 3750 vehicles per hour, the tunnel authorities tried using three lanes in one direction at peak hours to assist in the tunnels flow.

Liverpool City Engineer introduced a signal control system on the tunnel approaches with special queuing sections for waiting vehicles and this was further improved by the construction of a flyover for local traffic crossing Byrom Street, which was the main access to the tunnel. During this period, the Borough Engineer of Birkenhead also was responsible for the construction of tunnel approach flyovers linked with a large storage area and in 1970, all toll facilities were moved to the Birkenhead side.

On the Liverpool side there is a large 'trumpet' junction to Scotland Road which was widened and improved in the late 1970s to urban motorway standard and which will provide free flow access southwards to the Liverpool Inner Ring Road. The approach roads are designed to allow a good traffic flow between Leeds Street and Scotland Road in Liverpool and the toll area in Wallasey. They access roads have a traffic capacity in excess of the tunnels and at peak times, the Wirral bound tunnel has two-way traffic to reduce any congestion at the tolls.

The final stages of both approaches are sunk below ground level with retaining walls on both sides. The depth and details of the retaining walls was dictated by the tunnel portal levels and the topography of the sites. The retaining walls in the entrance to the tunnels reduced noise and were better than elevated roadways, which would be a blight on the overall skyline and surrounding buildings.

The Liverpool Approach road has a gradient of 3.33% over a distance of approximately 3880 foot (1182.6 metres) with a 1300 foot (396 metres) of the cut-and-cover tunnel cutting at the. Vehicles leaving the tunnel via the west slip road must travel approximately 2800 foot (853.4 metres) from the before they can use the full width between the twin tunnels to provide an additional climbing lane for heavy traffic between the portal and the west slip road.

The tunnel approach and exit layout is governed by the limitations of space controlled on the eastern side by the realignment of Great Homer Street. This is a principal arterial route to the city with a dual 24-foot (7.3 metres) wide carriageway designed for a speed of 30 mile/hour at-grade intersections. The tunnel approach 'trumpet' intersection with Scotland Road, was widened to provide a 36 foot (10.97 metres) wide dual carriageway. The slip roads are 24 foot (7.3 metres) wide with an 8 foot (2.4 metres) hard shoulders, 3 foot (914 mm) paved strip allowing for a minimum separation on the offside of 5 foot (1.52 metres). The space between retaining walls in the dual carriageway trumpet section has been laid out with 11 foot (3.35 metres) and 5 foot (1.52 metres) hard shoulders for in-going traffic respectively. The 11 foot (3.35 metres) hard shoulder has been designed as a third lane for tidal flow working after tunnel duplication.

The design of the bridges has been dictated both by the ground conditions on each site and by the necessity for construction of two or more stages in order to maintain statutory services and traffic during the whole period. Bridges over the Liverpool cutting have been constructed as two-pinned reinforced concrete portals because the horizontal thrusts at the bases are readily accommodated.

Given that the tunnel approach forms a large area at street level, provision had to be allowed for pedestrian access to the various parts of the outer city areas. An allowance was made to form four subways and two pedestrian bridges. At Limekiln Lane, the bridge is a pedestrian only bridge. At Scotland Road and Great Homer Street, Fox Street is carried across the approach cutting by a bridge leading into the area enclosed by the loop of the junction currently occupied by a large Bestway Warehouse. Vauxhall Road, a radial route to the city, is carried over the cut-and cover as well as the external sun visor section.

As both sides of the river are busy seaports, provision had to be allowed for large and unusual loads that needed to travel through the tunnels. An inspection bay is situated along the main access road leading to the tunnel (under the Scotland Road Bridge) to accommodate an observation booth and a 20 foot (6.0 metres) wide inspection bay. In addition, a 16 foot (4.87 metres) wide escape route with a 10% gradient is provided for vehicles that are prohibited by the by-laws or have entered the tunnel approach roads by mistake.

The Wallasey Approach road has a gradient of 4% over a distance of approximately 2743 foot (863 metres) with an 1100 foot (335 metres) of the cut-and-cover tunnel cutting and the Toll Area. The local properties that were in close proximity which were in a dilapidated state were demolished to allow sloping banks to be cut in the clay, some 37 feet (11.2 metres) deep, overlying the sandstone. The sloping banks had trimmed vertical and faced with reinforced concrete to prevent erosion and included 75 foot (22.8 metres) high retaining walls. The carriageway is 60 foot (18.2 metres) wide, two 4 feet 6 inch (1.37 metres) hard shoulders with a 3 foot (914mm) paved areas.

This area accommodates five 12 foot (3.65 metres) lanes, so as to provide an additional climbing lane for heavy traffic. The steeper tunnel gradient on this side of the river is partly compensated for by the availability of this crawler lane from a lower level. The toll area is approximately 1000 foot (304.8 metres) long and 217 foot (66 metres) wide, providing 13 lanes and 12 booths. The middle six booths can be operated in either direction to provide a maximum of nine lanes in the direction of major flow. The normal clear width is 11 foot 6 inch (42 metres) between booths and 8 foot 3 inch (30 metres) between the kerbs.

At the edges of the toll booths there are wider areas for the likes of articulated vehicles and motorcycles. The motorcycles have their own narrow lanes at the either end of the tollbooths to allow them to pass freely through in either direction. The articulated vehicles have lanes of 9 foot 9 inch (35.6 metres) and 20 foot (6 metres) between kerbs. Immediately south of the toll area is Gorsey Lane intersection, which takes the form of a conventional roundabout on the realigned and improved Gorsey Lane, with the tunnel approach road passing underneath. Between Gorsey Lane and Mill Lane, the works are generally within the limits of the old railway land.

By removing the earth embankments and replacing them with retaining walls to a minimum width of 75 feet (22.8 metres) allowed for the accommodation for a dual 24-foot (7.3 metres) carriageway with 3-foot (914 mm) paved strips, 8 foot (914 mm) hard shoulders and 5 foot (1.5 metres) wide central reserve. Between Mill Lane and Breck Road, fluted walls have been constructed where the cutting is exposed with naturally occurring sandstone, and the 3-foot (914mm) paved strips, which are widened to 5-foot (1.5 metres) verges.

The alignment of the former railway allowed for a tunnel approach road design to allow a speed limit of 40 miles / hour. At the Poulton Road Bridge, this 40 miles / hour limit is more appropriate as it houses a curve with a radius of 833 feet (2.5 metres) and this is considered a reasonable speed limit. The main consideration for this is that the radius of the curve and reduction of speed from the M53's 70 miles / hour which sees the vehicles reduce speed to zero at the toll Booths.

As with the Liverpool side, allowance was given to local traffic and pedestrians across the tunnel approaches. Seven original brick bridges had to be replaced with larger bridges to allow the existing traffic to travel across the approaches. In four instances, the cutting had to be deepened to allow for the increased headroom need for the tunnel traffic. A short length of piled carriageway has been constructed where the route passes close a gas holder with structures and foundations below road level.

The architectural emphasis is vertical, the exposed wall faces having corrugated ribbing with 6inch x 6inch (150 x 150 mm) indentations at 12-inch (304 mm) centres. In order not to detract from the aesthetic value, it was required that the horizontal joints should be as unobtrusive as possible, and this posed a problem. The design of the bridges in Wallasey, at Wheatland Lane, Oakdale Road, Gorsey Lane intersection and Poulton Road are of in situ reinforced concrete slab construction. In comparison, the 83-foot (25.29 metres) span of the Mill Lane cellular reinforced concrete box construction bridge carries main drains, which require a depth of 7 foot 2 inches (2.18 metres). The bridge was built in three parts by first forming the footpaths, incorporating service ducts on both sides, and then, after diverting the services and demolishing the existing bridge the roadway was completed.

The two remaining bridges, Breck Road and Docks Links road are constructed of welded steel plate girders and high tensile steel box girders, respectively with both having a composite reinforced concrete deck. The bridges carrying the M53 Motorway over West Kirby railway line and the Birket watercourse, together with Leasowe Road flyover on the Wallasey connection are largely in situ reinforced concrete bridges, but in each case inverted precast T beams have been used on the main spans where centring would have been expensive.

The cut-and-cover structure method used on the Liverpool end forming part of the approach extends over a length of 800 feet (243.8 metres) and has a width at road level varying from 103.5 feet (31.54 metres) to 86 feet (26 metres). The retaining walls in boulder clay, although, founded on sandstone a little below road level, had to sustain a continuous series of heavy horizontal loads. Cantilevered walls went to a height of 58 feet (17.6 metres) with propped cantilever struts spanning 100 feet (30.48 metres). Given this and the possible problems associated with the proposals an arched strut, was developed into a continuous arch, and was the solution adopted, with backfilling over the arch.

Beyond this cut-and-cover section, the rock level rose to levels that would allow a cantilevered retaining wall to be designed along with some sections having battered slopes formed along their length. On the Wallasey side, retaining walls of similar design were used throughout. The exposed face was shaped with near vertical rectangular corrugations about 6 inches wide by 6 inches (150 x 150mm) deep, battered back from the vertical at 1 in 20.

Wallasey Tunnel Dates of Specific Milestone Events

1963
02nd November – Committee decide on a new tunnel
1964
15th February
Seven local authorities discuss idea
27th November Tunnel Bill before Parliament
23rd December
Wallasey and Liverpool electros approve tunnel
1965
06th August
Tunnel receives Royal Ascent
1966
13th January
Lord Mayor of Liverpool Alderman David Cowley starts drill at 11.35
Mayor of Wallasey Alderman C. S. Tomkins starts drill at 12.05
10th March
Vertical Shaft at Liverpool 80 feet (24.38 metres) and Wallasey at 60 feet (18.28 metres)
01st April
Demolition of houses for approach roads
31st May
Pump breakdown
14th October
Pilot Tunnel partially flooded after fire at pumps.
Workmen fought to hold 750,000 gallons of rising water
18th October
Normal work resumes
15th December
Pilot shafts 700 feet (213 metres) apart

1967
17th January
Breakthrough
31st October
Engineers arm mole cutters at Wallasey
02nd December
Mole starts work
1968
24th June
Day and night shifts drive mole 212 feet 6" (64.77 metres) which was claimed as a new World record
19th November
Major flaw in riverbed halts mole
31st December
Mole starts again
1969
27th February
Water and Silt causes mole thrust to stop work
17th March
Underground repairs to mole set up an engineering feat with the bearing
22nd May
Mole starts upward climb
16th October
Civic Leaders walk (or rater paddle) through from Liverpool to Seacombe and meet the mole on the way through
1970
04th March
Breakthrough at Liverpool

Once the Mole had finished its work, it was dismantled and put into storage on Waterloo Dock near Great Howarth Street, close to the tunnel dock exit. Assessors looked at the viability of the machinery and possible restoration. The costs of this would have been approximately £100,000 (£1,436,870.00) or scrapping it which had a value of £5,000 (£71,843.50). The only other option was to purchase the Mole for £15,000 (£215,530.50) and make a monument out of it.

In an article in the Liverpool Echo in May of 1972, there was a discussion as to the viability of using the mole for a monument (Mott Hay Anderson) or even giving the mole to an under developed country. The mole was the holder of several world records during the construction of the Kingsway tunnel one of which was a drive on tunnel two for a length of 7,400 feet, (2,255.22 metres) in 14 months.

Wallasey Tunnel Statistics
Portal to Portal Approximately 1.5 miles
Length including approach roads Approximately 5.5 miles
Width of river 1000 yards
Diameter of pilot hole 12 inches (horse shoe) (305mm)
Diameter of main tunnel 30 inches (Circular) (762mm)
Depth below river 20 feet (6.1 metres) minimum or 40 feet (12.2 metres) average
Gradient 1:25 max
Road width 2 x 12 feet (3.65 metres)
Headroom 16 feet 6 inches (5 metres)
Lining: Mostly concrete segments / cast iron

Ventilation: One each side

The construction cost of the tunnel project including those approaches exclusively used by tunnel traffic will amount to about £29 million, (£357,978,900.00) with an addition of £2.8 million (£34,563,480.00) for land and rehousing.

Ancillary road works for Wallasey County Borough Council amount to £2.7 million (£33,329,070.00) and for Liverpool City Council £0.4 million (£4,937,640.00). The motorway link across Bidston Moss for the Department of the Environment cost about £3.3 million (£40,735,530.00).

The tunnel costs may be allocated as follows:

Costs are in Millions of pounds (£)
(Today's equivalent figures shown in brackets)

Twin tunnel construction between portals	13.0 (£160,473,300)
Cut and over tunnel	1.8 (£22,219,380)
Tunnel paint finishing's and equipment	5.3 (£65,423,730)
Ventilation and administrative buildings	1.2 (£14,812,920)
Approach roads	7.6 (£93,815,160)
	28.9 (£356,744,490)

The scheme provides an interesting cost comparison for four different types of construction to motorway standard.

The figures are all for the provision of dual two-lane carriageway.

Costs are in million pounds per mile of tunnel

Order twin tunnels excluding ventilation works and services	9.5 (£117,268,950)
Cut and over tunnel (Liverpool) excluding lighting and adjusted from the actual width of about 100 ft to 75 feet	8.0 (£98,752,800)
Open cut approaches (Wallasey) with retaining walls average depth 20 ft wide, 75 feet including overbridges	
Elevated viaduct over Bidston moss, width 60 ft deep piled foundations (approx. 200 ft on average)	3.5 (£43,204,350)
The open cut approaches at Liverpool cannot be brought simply into the comparison because of the complex junction.	

The tunnel was driven mainly in Bunter sandstone by machine. The rock did not respond satisfactory to treatment by injection. Copious inflows of saline water affected plant, hampered work, and often turned the spoil at the face to slurry. Poor rock in the roof produced material very difficult to handle, particularly when mixed with the slurry. The treatment of a major fault, the changing of the main bearing of the TBM and the repair of severe damage to the machine caused by breakage, of a pinion, all near mid-river, provided some interesting features in driving Tunnel 2A. Experience gained, added to the skills required on that drive, enabled Tunnel 2B to be driven at much greater speed.

Estimates of the overall costs of the works and of forecasts of the quarterly expenditures for their completion were prepared, both being apportioned in accordance with the requirements of the several Clients. These estimates and forecasts were reviewed each quarter and adjusted when necessary. They gave the Mersey Tunnel Joint Committee the information necessary for applying for loan sanctions pursuant to the Act of 1965, and enabled the local authorities to make timely applications for grants for road works. The Clients were thus provided with a reasonable measure of financial control and could budget not only for final expenditure but also for interim payments to contractors.

The Resident Engineers allocated the amount of each interim measurement on an agreed basis of apportionment. A summary of the allocations of the corresponding monthly certificates was submitted to the Clients to enable the Mersey Tunnel Joint Committee or Department of the Environment to recover allocated costs under the terms of their agency agreements with the respective authorities.

Wallasey Tunnel 2B

A Contract for the second (Wallasey to Liverpool) tunnel was awarded to Nuttall Atkinson & Co to be started on 1 December, 1969. The axis of the pilot tunnel would be 1 feet 6 inches (457mm) below that of the main tunnel. This would see the tunnel cut via an 11 feet diameter tunnelling machine was designed specifically for the pilot drive. The Mini-TBM has steering that has been designed to deal with the highly abrasive rock and arising. In addition to this, it also has a series of suction tubes through which water could be drawn off from just behind the main cutter head, along with a central Numec belt conveyor to handle the very difficult material.

Despite the Mini-TBM having a good digging rate, it failed when it entered the wet ground some 350 feet from the portal, where the water table was reached. The main reason for this failure was the large quantity of water and slurry produced during the digging process. This 'Mini-TBM' started work on 23 March 1970 and attempts to work with it were finally cancelled on 19 May. The tunnel was further excavated by hand driving a 10 feet 6 inches x 10 feet 6 inches (3.2 x 3.2 metres) heading started at one face on 27 May. The under-river drive was started on 22 June with the breakthrough at mid-river 26 January 1971. In all a total of 6442 feet (1,963.5 metres) of pilot tunnel were constructed.

Upon completion of Tunnel 2A, the main chassis of the TBM was transported across the Mersey on 2 July 1970, using the Dock Board's 100-ton Atlas floating crane. The TBM was then transported to the Wallasey site and reconstructed. The works to rebuild the TBM were completed in the best part by mid-October, when the machine was moved up to the Wallasey Portal face. After testing, the TBM was complete enough for the next tunnel drive to start on 11th November.

Learning from the lessons of the first tunnel, the whole process was thought through and revised for tunnel 2B.

Some of the main areas that were seen in need of a rethink are as follows:
- Method of dealing with the main fault.

Working from the pilot tunnel a roof of heavily reinforced concrete beams formed above the main tunnel alignment as in Tunnel 2A, in advance of the arrival of the TBM and without affecting its progress.
- Protection of the main bearing of the TBM.

The system devised and installed at the time of the bearing collapse on the first drive was improved and reused. Worn metal faces were built up. Greasing was enormously increased.
- Material and system for grouting the annular space around the lining.

The grouting plant was repositioned much closer to the face than before, indeed as close as the redesign would allow. Long-chain polymer were introduced into the grout.
- Muck handling system on the TBM and back through the tunnel and onto disposal.

This was redesigned to cope with slurry in large quantities of rock which might block the conveyors and a mixture of both materials, while still maintaining the required work rate. A special travelling mechanical discharger from the conveyor system was provided for muck stacking in the bin outside the tunnel, because of the restricted area available when constructing the duplication.

The redesigned TBM and back-up equipment worked well, and completed more than half of the tunnel in less than seven months. On 12th July 1971, the teeth of the main ring gear were seen to be scuffed. Many were made to sustain the damage but the inevitable was to happen. The machine was stopped on 10 September to change the buckets and arrange a reversal of the direction of rotation of the cutter head. After restarting on 28th September, very heavy wear occurred on the main bearing, and it broke down two days later. One roller path had become largely detached from the bed metal. Replacement of the bearing was completed in three weeks and the drive was resumed until the tunnels completion within 14 months.

Finishes

The steel inner face to the structural lining has been protected and decorated. This was followed by blast cleaning which was followed by a zinc-rich epoxy primer and a hot applied solvent free epoxy coating. No other finish has been applied to the crown of the tunnel which is dark in colour but the sides above the dado panels received a further coat of white polyurethane paint.

Following extensive exposure trials in the Queensway Tunnel this paint system was designed for long life in tunnel conditions, which include regular washing and scrubbing by tunnel cleaning vehicles. For the sections of tunnel lined with cast iron segments, a similar finished appearance is provided by the installation of a false lining made up of coated steel sheets bonded to asbestos cement and fixed to a steel framework secured to the cast iron lining. The approximate cost of false lining to the cast iron segments was £20.50/sq yard as compared with £5.75/sq yard for the grit blasting and painting of the concrete/steel segments.

Along each side of the tunnel cable runs are covered by dado panels 7 feet (2.1 metres) high formed from vitreous enamelled steel sheets bonded to asbestos cement. Tunnel walkways on both sides for inspection and maintenance were assembled from L-shaped units precast in 12 feet (3.65 metres) lengths. They incorporate manhole access to a services duct, and ventilation slots to the roadway white ceramic tiles facing the roadway were incorporated during manufacture.

The arched cut-and-cover section between the Liverpool Portal and the open cut is faced with painted concrete slabs covering the large bored piles, while the arch soffit is of cast concrete. There are six light wells which provides gradation of daylight in the outer half of the section and eight extract vent shafts. The whole provides an uninterrupted roofed space nearly two acres in extent.

The arch form is extended for a further 200 feet (60.96 metres) by a sun visor system in which reinforced concrete ribs at 20 feet (6 metres) centres and matching the profile of the adjacent arch carry precast concrete sun visors of honeycomb pattern to shade the tunnel approach roadway from direct sunlight and to graduate the change from full daylight to artificial light. Similar sun visors are provided at the Wallasey Portal, when the duplicate tunnel is driven.

Lighting

The tunnel has lighting throughout its length by a continuous central line of twin lamp fluorescent fittings in the crown of the tunnel. The lighting units are corrosion resistant, hose proof and are designed to direct a high proportion of the light from one or two 8 feet long 85 W tubes to the kerbs and sides of the tunnel.

Maximum light in the tunnel is 160 lux, (unit of light measurement) increased at the entrances to 1100 lux by additional lines of fittings. Photo-electric cells regulate the light at the entrances in accordance with external conditions and at night cut the main tunnel lighting to 75 lux by switching out one tube in each fitting. The portal lighting is correspondingly cut to 60 lux. In the Liverpool arched portal, the fluorescent tubes are supplemented by nearly 300 mercury discharge lamps.

This lighting was updated in 1981 from the amber fluorescent tubes on the walls of the tunnel. The reason for the change is that the old lighting was inefficient and ineffective and causes a flicker effect on vehicle windscreens. This reflection could see the onset of an epileptic fit in susceptible people.

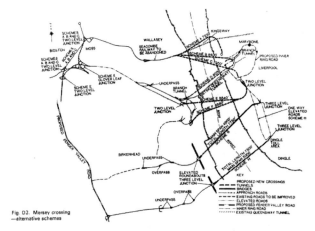

Tunnel Approach Route Planner (Cairncross and Jones)

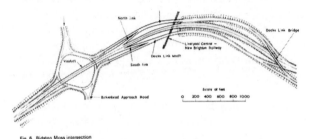

Bidston Moss Link Road
(Cairncross and Jones)

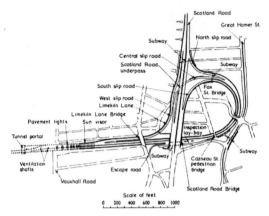

Liverpool Exit (Cairncross and Jones)

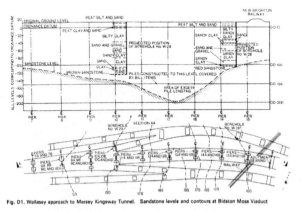

Wallasey Approach Sandstone Levels at Viaduct (Cairncross and Jones)

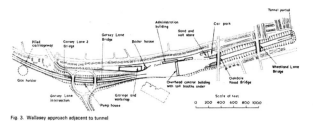

Wallasey Tunnel Approach Adjacent to Tunnel (Cairncross and Jones)

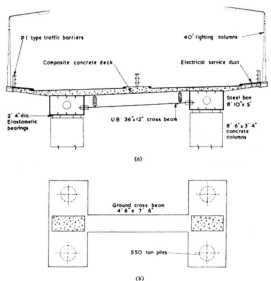

Typical Section through at Bidston Moss (Cairncross and Jones)

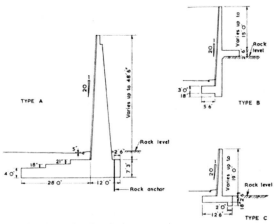

Typical Wallasey Approach Retaining Wall Section (Cairncross and Jones)

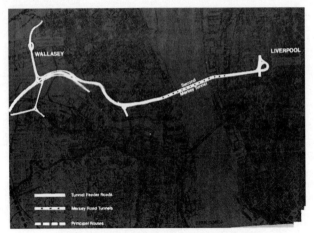

Proposed Route (Birkenhead Library)

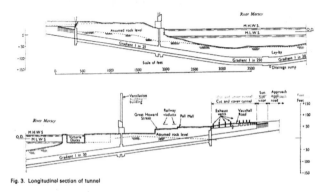

Longitudinal Tunnel Section (Megaw Brown)

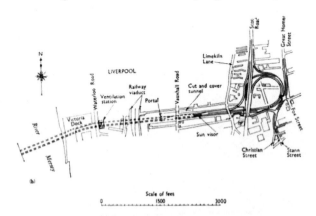

Liverpool Tunnel Exit (Megaw Brown)

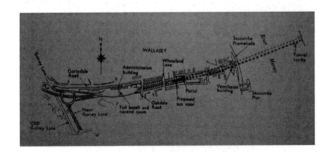

Wallasey Tunnel Approach (Megaw Brown)
The following are General Construction Pictures of Kingsway Tunnel
Some of which have specific titles, others provide an insight into the conditions within
the tunnel during its construction (Birkenhead Library)

Cement Pallets on Gantry

Concreting Pilot Tunnel fault Nov 68

General Tunnel Section

Slurry Invert Oct 68

Slurry Removal

Surveyor

Working Conditions

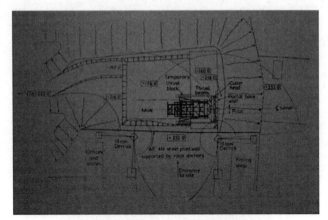

Tunnel Segment Site Layout (Megaw Brown)

Tunnel Segments Awaiting Delivery to Site (Birkenhead library)

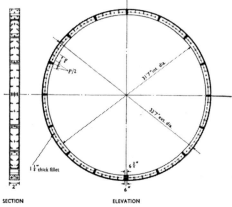

Fig. 6. Cast iron lining for Liverpool hand drive

Lining for Liverpool Hand Drive (Megaw Brown)

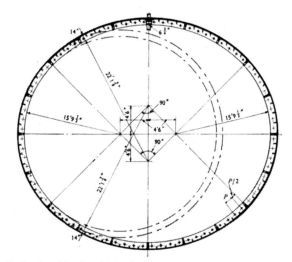

Fig. 7. Cast iron lining for mid-river lay-by

Lining for Mid River Lay-By (Megaw Brown)

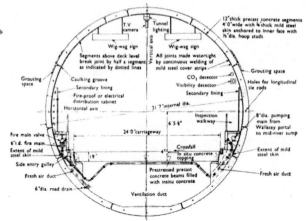

Section through Pre-Cast Concrete Lined Tunnel (Megaw Brown)

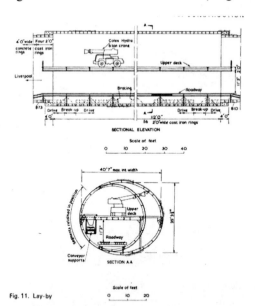

Tunnel 2B Construction of Lay-By (McKenzie and Dodd)

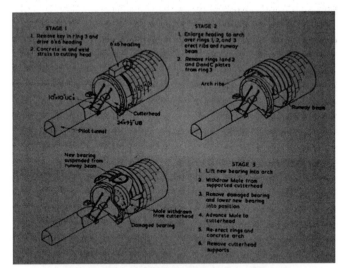

Mole Bearing Change (Birkenhead Library)

Mini Mole Machine (Birkenhead Library)

Main Mole Machine Cutter Head (Birkenhead Library)

Main Mole Machine (Birkenhead Library)

Break Through (Birkenhead Library)

Dany Williams (Driver) and Bill Jobin (Shift Boss)
Who made the final push 05-03-1970 (Liverpool Echo)

Mole Recovery Shaft (Birkenhead Library)

Liverpool Sun visor under Construction (Birkenhead Library)

Placing Walkway Segments (Birkenhead Library)

Managers Survey the Finished Tunnel (Birkenhead Library)

Second Tunnel Drive Team (Birkenhead Library)

Chapter 13
Bidston Moss Piling

The requirements for a junction at Bidston Moss presented a complex design problem involving traffic to and from the tunnel, the M53 Motorway connection, and the Wallasey, Birkenhead docks connection, with a requirement for diversion of traffic in emergencies between the Queensway and Kingsway tunnels. In the layout adopted the ground level roundabout handles local traffic, while all tunnel and dock traffic takes place on the Docks Link viaducts which have three-lane-wide sections about 1000 feet (304.8 metres) long allowing traffic merging to take place at up to 45 mile/h.

It was forecasted that by 1990 there will be 3330 vehicles/h on these merging lengths, while the through route carries slightly under 2000 vehicles/h in the direction of major flow. This allowed for a dual two lane carriageways to be provided for the traffic requirements. On the through route, a hard shoulder have been provided with the exception of the viaduct section where they have been omitted as the cost is not justified for the traffic volumes forecast.

The Motorway (at the time) was a standard Department of the Environment width, being dual three lanes with hard shoulders. Elsewhere the carriageway is 24 feet (7.3 metres) wide with 10 feet (3 metres) wide verges. The roundabout at Bidston Moss has merging widths generally 40 feet (12.19 metres) wide and 250 feet (76.2 metres) long with the exception of the NE leg which is 50 feet (15.2 metres) wide. The forecast maximum flow through this section was over 4000 vehicles/h in 1990 with a reserve capacity of just under 10%.

The earthworks problem revolved about the scarcity of fill at the north end of the Wirral where the Bidston Moss works between Breck Road Bridge and Fender Lane required 1.3 million cubic yards (993,921.31 cubic metres) for embankments.

At Liverpool, the cutting excavation produced about one third of the Bidston requirements and this was transported through the existing tunnel, mainly at night, to form pre-contract embankments. At Wallasey, the excavation from the tunnel and the disused railway cutting provided another third, which was mainly used in embankments between Breck Road and the New Brighton railway. Although, minor quantities were imported, the remaining third of the total was obtained from local borrow pits. Clay was excavated in a suitable area of the Moss, thus providing valuable tipping space for refuse disposal. Dune sand was used for filling below ground in one section.

There is a particularly bad section of ground with peat up to 20 feet (6 metres) deep to the west of the New Brighton railway which itself stands on a low embankment. Embankments 30 feet (9 metres) high would have been required to cross the railway and the intersection roundabout: consequently, because of the scarcity of suitable fill and the low load-bearing capacity of the underlying strata, it was decided to construct this section of height approach roads on a viaduct. In addition to the Bidston Viaduct, seventeen bridges were built on the approaches. The Docks Link Bridge is 440 feet

(134 metres) long continuous over four spans, there are four three-span bridges, 103 feet (39.6 metres) to 180 feet (54.8 metres) long, and the remainder have single spans of 51 feet (15.5 metres) to 127 feet (38.7 metres).

The main through route, 2400 feet (731.5 metres) long, is carried on a 16-span structure. The spans vary from 126 feet (38.4 metres) to 159 feet (48.4 metres) and are continuous over three, four or five piers between slung spans. The north and south Docks Link roads are carried on similar structures of nine spans and eight spans respectively with span lengths varying from 98 feet (29.87 metres) to 159 feet (48.4 metres). The connecting spurs, the north and south links between the through route and the Dock Link each consist of two independent spans of 130 feet (39.6 metres). The levels, thicknesses and bearing capacities of the layers of sand, gravel and clay underlying the peat and silt are so variable and uncertain that the viaduct is founded on 134 piles, 4 feet (1.21 metres) in diameter, bored down to the sandstone. The acceptable stress on the rock was 45 ton/sq ft giving a working load of 550 tons per pile.

The water table is immediately below ground level and the sides of the holes required the support of a 6% bentonite suspension during boring. A Calweld 150 CH rig mounted on a RB 38 was employed for drilling. The buckets used, except in the stiff clay layers was a 42-inch (1.06 metres) diameters auger was followed by a 48 inch (1.2 metres) diameter auger to ream out to the full diameter.

The concrete for the piles was specified to contain not less than 700 lb/cu yd (415.3 kg/m3) of sulphate resisting cement and to achieve a minimum strength of 4200 lb/sq inch (295.3 kg/cm3) at 28 days. A workable mix for placing through a 10-inch (254mm) diameter tremie tube was obtained with a water: cement ration of 0.47, giving slump tests of 5.7 in (144mm).

Exploratory boreholes on the line of the viaduct taken prior to construction indicated a maximum depth to rock of 145 feet (44.1 metres) and in the Bill of Quantities provision was made for boring piles to 165 feet (50.2 metres) below ground. Unfortunately, the siting of the boreholes was such that two adjacent holes came on either side of a buried channel in the rock some 500 feet (152 metres) wide.

The Contractor's rig was capable of reaching a depth of 170 feet (51.8 metres) below ground and completing 97 piles. By making a special extension piece to the 'Kelly' an additional 25 feet (7.6 metres) was obtained to reach a depth of 195 feet (59.4 metres) and complete 17 of the remaining piles. In order to bore the last 20 piles, a shallow cutting, 16 feet (4.87 metres) deep down to pile cap level, was excavated and by working the plant from this level, the deepest pile, 210 feet (64 metres) below ground.

Incremental loading tests to 550 and 825 tons were carried out on three piles of 97 feet (29.5 metres), 176 feet (53.6 metres), and 200 feet (60.9 metres) long. Strain gauges were installed in the 200 feet (60.9 metres) pile in order to obtain load/depth distribution during the testing cycle. Comparison of computed and observed load/settlement curves indicates a high level of load transfer, particularly noticeable on the two deepest piles, where this is attributed to over break in cohesion less sand layers near the surface.

In order to keep these expensive foundations to a minimum the spans consist of steel box girders varying from 5 feet (1.5 metres) to 14 feet 10 inches (4.5 metres) wide and 5 feet (1.5 metres) deep, acting compositely with reinforced concrete decks. The boxes were fabricated in lengths of about 40 feet (12 metres), were erected on the concrete piers were joined by friction grip bolts. When each set of continuous spans had been completed, the deck concrete was placed in sequence with the pours in the tension zones over the piers last. The walls are 8 feet diameter contiguous bored piles keyed into the rock at the bottom.

The sequence of construction was that the area was excavated to arch springing level 28–40 feet (8.5–12 metres) below ground and from this level, the piles were bored and concrete filled. The dumpling between the piles could not be removed until the thrust from the arch had been developed by the overburden loading. The successive stages of dumpling excavation and filling over the arch were carefully regulated.

During 1964, 14 boreholes were sunk on the Liverpool approach and 63 on the Wallasey and Bidston Moss Approach. At the Bidston Moss approach, one borehole showed the sandstone, overlain by glacial and alluvial strata to a depth of 149 feet (45.4 metres). A further six boreholes were sunk in the River Mersey to a maximum depth of 100 feet (30.48 metres) into rock. This information along with various other surveys (Echo Sounder and Sparker Survey) was to ascertain the depth of rock and other material that was making up the ground for the proposed tunnel. It was also to see if there were any possible faults or weak areas that would become dangerous during the tunnels construction.

Bidston Moss Viaduct is a 730-metre-long box girder bridge constructed in 1970 and consists of a 16 span structure carried on 34 piers which carries the M53 over the A554 roundabout and the Birkenhead to New Brighton railway, and is jointly owned by the Highways Agency, Wirral Borough Council and Mersey Travel (Owners of the tunnel). It is used by over 50,000 vehicles daily including 3,000 heavy goods vehicles.

The box girder styled viaduct, was originally built at Cammell Laird shipyard in Birkenhead, and traditional shipyard skills of working with steel plates were used on the 200 plates. At a total weight of 10 tonnes, allows heavy vehicles to use the viaduct again after a substantial refit just after the 2006 bridge collapse in Australia.

During the exploratory works to enable the construction to be designed, the ground conditions were found to be varying and consisted of sand, clay and overlying sandstone. The contractor came across the water table at 5 feet 6 inches (1.7 metres) and during the boring process; this was found to vary from 10 to 33 feet (3 to 10 metres). When the exploratory pile linings were withdrawn the water pressuring within the linings was found to rise within 5 feet (1.5 metres) off the surface.

This preliminary work allowed the engineers to calculate the best position for the piles to ensure the stability of the structure. Not too many years before the construction of the viaduct, the ground conditions would have made it impossible to use, but with the onset of the continuous steel casing for support, the site has now become a viable option with the onset of modern techniques. It was decided to pile into the base rock and these ultimately lead to the use of the 200-foot (61 metre) piles on the site, which at the time was the largest pile in Europe.

When the engineers were considering a continuous steel casing within a pile, they have to ensure that the steel casing acts as though it is a single piece of light gauge steel and continue down to good ground at the piles required depth. This final destination may be increased by the use of a mechanical machinery to ensure the bottom of the pile is protected against the water table.

Once the initial pile had been drilled, a reamer was used to give the hole a slightly oversized diameter. The steel casing with an internal diameter of 4 feet 3 inches (1.3 metres) and 25 feet (7.6 metres) long. Drilling then continued to the required depth using bentonite and a standard 48-inch (1.2 metre) diameter bucket. The bentonite level was maintained between 15–20 feet (4.5–6 metres) below ground level. To enable the site to progress, it was necessary to install a complete bentonite mixing, settling and recovery plant on the site. This made the whole process a complicated one with a large site to facilitate for men and materials.

Bentonite is an absorbent aluminium phyllosilicate, essentially impure clay consisting mostly of montmorillonite. Much of bentonite's usefulness in the drilling and geotechnical engineering industry comes from its unique rheological properties. Relatively small quantities of bentonite suspended in water form a viscous, shear thinning material. At high enough concentrations, bentonite suspensions begin to take on the characteristics of a gel.

For these reasons, it is a common component of drilling mud used to curtail drilling fluid invasion by its propensity for aiding in the formation of mud cake. Bentonite can be used in cement, adhesives.

It was found that for very stiff clay at around 80–120 feet (25 to 36 metres), augers would be used and the best process would be seen by drilling 42 inches (1.07 metres) in diameter and then reaming the pile hole out to the required 48 inches (1.22 metres).

On completion of drilling, the bottom of the hole would be cleaned out and the bentonite checked to see what level of contamination there was. If the bentonite levels had increased, the bentonite would be pumped out and transferred to the settling tanks and replaced with fresh bentonite.

The next stage was the placing of the steel reinforcement and concrete in the pile hole. The upper 50 feet (15 metres) of the pile are steel reinforced cages of 1 inch (25 mm) diameter bars. Spacers on the sides ensure that the cage stays central within the hole as it descends and welded extensions at the top, allow it to be suspended at ground level during concreting. The concrete used in the pile holes gave a compressive strength of 4200 lb per square inch (300 kg per sq centimetre) after 28 days. The piles were concreted through a 10 inch (25 cm) diameter pipe which reached the bottom of the hole.

In accordance with the normal process, a bag containing gravel was put into the top of the pipe to form a plug. The first charge of concrete forced the plug down and out of the pipe at the bottom of the pile hole. This standard process allows the bag to act as a piston to carry the bentonite and ensuring that the concrete was not contaminated as it passed down the pipe. The displaced bentonite was pumped away into the recovery tanks. As the concrete filled the hole, the pipe was raised and shortened to allow the concrete to flow. The bottom of the pipe was always kept 10–15 feet (3 to 4.5 metres) below the surface to retain the seal.

The continuous concreting of the piles took a number of hours. For a 200 foot (61 metre) pile it was around three hours and contains around 90 cubic yards (70 cubic metres) of concrete. The working load of each pile is around 550 tons

In June of 2006, it was being widely reported that the Viaduct was to be demolished. The local paper the Liverpool Daily Post and Daily Echo ran an article on the plans by the then Highways Agency to demolish and rebuild the troubled motorway link. In the article, it stated that the problems started when a bridge of a similar design collapsed and killed hundreds in Darwin, Australia. The catastrophic collapse was of the part-constructed Westgate box-girder Bridge in Melbourne (Australia), which stopped similar projects all over the world as engineers worked out what had gone wrong and whether a repeat was likely. The review was also influenced by the failure of the A477 Milford Haven Bridge over the River Cleddau, also during construction.

After this disaster, the M53 Bridge has since had continual strengthening and concrete protection works carried out on it. This sparked a local debate on the future of the structure and its constant cost of continual maintenance. The original problems were of undersized welds in the box girders compared to the welds shown on the as-built drawings – hence the emergency weight restrictions being put into place years earlier.

If the structure was to be demolished, then consideration would have to be given to the Merseyrail line below. Although, this has nothing to do with the weakness of the bridge, if and when demolition goes ahead, the vibrations from passing trains may affect the bridge foundations to vibrate and weaken. Most of this land is re-claimed. Previously it was marshland but when the Wallasey Tunnel was constructed, the waste was tipped on what was soon to be a pretty major motorway junction

The article stated: *The Highways Agency has ordered a full investigation after millions of pounds and years of attempts to shore up the existing structure have failed to solve its problems.*

Officials from the Agency last night admitted demolition was one of the options they were considering and one source told the Liverpool Daily Post and Echo that the bridge would probably have to come down.

Strengthening work has just been completed to allow HGVs over the viaduct bridge; although, the lorries will still be restricted to one lane.

Charles McKeown, vice-chairman of the Federation of Small Businesses in Wirral, said,

"That the traffic problems caused by demolishing and rebuilding the viaduct would have a massive impact on Wirral firms who depend on the link to Liverpool. It is the main route to the Wallasey (Kingsway) Tunnel under the River Mersey and is used by more than 50,000 vehicles daily including 3,000 heavy goods vehicles."

In January 2005, the decision was taken to ban heavy goods vehicles completely from the viaduct as a safety measure.

Heavy vehicles travelling to and from the tunnel along the M53 were diverted via Junction 1 around the roundabout and then back onto the motorway via the parallel slip road.

Mr McKeown, who works in the construction industry, said

"The ban on HGVs had caused enough problems for drivers and commuters."

He stated further that:

"It created havoc, with traffic backing up all the way to Wallasey and Birkenhead, especially in the mornings and evenings. It's a key link between Wirral and Liverpool and most people choose to use the Wallasey tunnel, unless they're coming from Birkenhead or the A41, because it's at the end of the motorway.

"If they were to send all the traffic along the slip roads and around the roundabout at B&Q it would be an absolute nightmare. It would create major problems for all businesses, small, medium and large."

The Highways Agency has said a full technical survey of the viaduct on the M53 is now being carried out to determine its long-term future; although, this is at an early stage.

One possibility is that a replacement bridge could be built alongside the existing structure and then traffic switched when work on the new viaduct was complete; although, the cost of this approach has been described as 'significant'.

A spokesman for the agency said:

"We are looking at the future of the viaduct and nothing has been ruled in or out. They are doing tests on whether there could be further strengthening, and we will do other tests on the structure to see how long a lifespan it has and the best way to take things forward. Last July, the £4m project to strengthen the viaduct, which is a box girder structure originally built at Cammell Laird shipyard in Birkenhead, was started."

Traditional shipyard skills of working with steel plates were used and more than 200 plates, with a total weight of 10 tonnes, have been bolted to the structure to allow heavy vehicles to use the viaduct again.

The Viaduct has had somewhat of a continuous history as far as repairs and maintenance is concerned. During 1998/99, the viaduct was strengthened to meet new legislation introducing 40 tonne vehicles onto the country's roads. However, during this work the need for further strengthening works was identified and as a safety precaution heavy goods vehicles were restricted to the inside lane of the main carriageway, a 3.5 tonne limit was imposed on the outside lane and the entry and exit slip roads reduced down to one lane. A further restriction in early 2005 reduced the capacity to cars only on the viaduct.

An initial strengthening contract carried out between November 2005 and April 2006 returned the viaduct to HGVs on the inside lane and a 3.5 tonne limit on the outside lane. Technical surveys, studies and assessments were completed and as a result it was proposed to carry out a multi-million-pound scheme to strengthen the existing structure while keeping disruption to the existing traffic to a minimum. Works include steelwork strengthening, painting (inside and outside the boxes), parapet refurbishment, new lighting, drainage improvements, limited resurfacing to slip roads and more extensive concrete repairs to the underside of the structure and piers.

Given this history and the prospect of demolition, the Highways Agency have opted for the strengthening and refurbishment of the bridge rather than the costly demolition, which would have been a huge inconvenience to the local road and tunnel network.

The works involved men working in extreme confined space environments where specialist training and skills were required to complete the strengthening operations. There were approximately 62 boxes to be fully strengthened, pier repairs, motorway deck replacement, lighting and electrics to be replaced. The scheme revolved around being designed for maintenance to ensure any future works can be complete quickly, efficiently and to a safe manner. The project was set-up with the workforce and public's safety in mind. With experience from two other box girder strengthening projects the project team was be able to deliver a safe and successful project for all parties involved in the scheme.

Tenders to bid to undertake the work were invited in June 2008 and the contract was awarded to Costain Limited in January 2009. The initial period of the contract and design phase commenced at the end of January 2009. Works involved site surveys, inspection, testing, detailed design of the strengthening works and some advance works in preparation for the second phase of the contract. On March 22nd 2010, Transport Minister Paul Clark announced that the funding for the second phase of the works had been approved ahead of programme.

The full strengthening works started in May 2010 and completed in September 2011, three months earlier than planned. Since then all weight restrictions and roadworks on the M53 have been removed and heavy goods vehicles have been able to travel in both lanes on either side of the motorway

Technical surveys, studies and assessments were completed with the studies concluding that online strengthening provided the best value solution to the viaduct problems. As a result a multi-million pound scheme to strengthen the existing structure, whilst keeping disruption to the existing traffic to a minimum, was developed.

Outline objectives for the scheme were:
- To restore the network to full capacity;
- To minimise traffic disruption on the network;
- To minimise any environmental impact;
- To ensure public safety.

A monitoring regime is in place to ensure the continued operational safety of the structure whilst strengthening proposals are being developed as part of the new strengthening scheme.

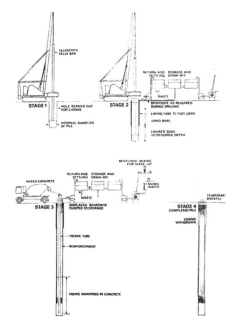

Typical Large Diameter Piles (Cairncross and Jones)

Chapter 14
Kingsway Tunnel Ventilation

Unlike the Queensway Tunnel, the second Mersey Crossing was to have one Ventilation building on each side of the river. One was to be a short distance from the Seacombe Ferry Terminal in Wallasey, and the other within Liverpool's old railway goods station in Waterloo.

The site of the Liverpool building has a large Costco wholesale building a short distance along the road, and the Old Waterloo Building is now flats. Both sites and buildings are easily distinguished due to their individual design unlike the Queensway ventilation buildings.

The Ventilation Buildings are positioned on either side of the river in a direct line with each other and as close to the river as possible. The Seacombe building has a guide along the promenade showing the line of the tunnel. The system was designed by Dr B. R. Pursall as a consultant to the committee. Each ventilation station draws fresh air into the tunnel via a horizontal shaft into the invert of the tunnel. Foul air is exhausted from the crown of the tunnel and discharged through the top of the ventilation stations. The towers discharge at a height of 162 feet (49.3 metres) on the Liverpool side and 185 feet (56.38 metres) on the Wirral (Seacombe) side.

The ventilation stations supply a balanced flow of air through the tunnel via a land and mid-river feed. Both sites initially had a dual purpose as they served as working sites for the pilot and main workings for the tunnel. The extract shaft is a central opening from the crown of the tunnel leading up to the fan which forms a rectangular aperture 28 feet x 20 feet (8.5 x 6 metres) in the crown of the tunnel. To enable this to be installed, segments were omitted over a width of five rings, to a 20 feet (6 metres) diameter circular shaft at ground level. It also has welded steel plate lining similar to that in the tunnel to aid the waterproofing process and is strengthened by a concrete strut.

The blowing shaft is 42 feet (12.8 metres) from the centre of the tunnel whose ducts are designed to carry the air as smoothly as possible from the circular fan opening at floor level into the invert of the tunnel. This circular aperture (17 feet 3 inches (5.25 metres) in diameter) leads into an octagon and then a rectangle (20 feet x 14 feet 3 in.) (6 x 4.3 metres) which is divided into two areas proportionate to the airflows required landwards and riverwards.

Each rectangular section sweeps round from vertical to horizontal axis and is further reshaped with minimal change of area and direction to lead as smoothly as possible into the tunnel invert and adapt to the shape necessary. On the Wallasey site, the shafts were sunk into sound rock, lined with reinforced concrete by bentonite slurry trenching and completed in the rock with minimal use of explosives. On the Liverpool site, timbered pits were sunk into the rock with similar rock excavation thereafter.

This complex system of large ducts connecting to the tunnel was designed in the knowledge that the rock would lend itself to the necessary techniques in contrast to other types of ground where construction problems would prevent the aerodynamic requirements being met in the same degree.

The blowing ducts connect horizontally into the invert of the main tunnel immediately below road level via the in situ reinforced concrete forming 44 feet (13.4 metres) long rectangular box for entry of each supply duct. Special provision has been made for the gas mains and electric cables to pass under the air ducts and avoid obstruction to the tunnels airflow. At the tunnel entrances, 32 feet (9.75 metres) length of tunnel accommodates the exhaust shaft in the crown

Semi-transverse ventilation was adopted as being an efficient and convenient for a tunnels circular section where the invert space below the road deck is available as a fresh air supply duct, and where there is ample space overhead for the flow of the exhaust air. This allows a maximum air supply of 200cu feet/min (5.66 cubic metres/min) of two-lane tunnel was adopted to meet the case of peak traffic filling the two lanes and ascending on a gradient of 1 in 25 or 1 in 30. As these conditions, could occur in either direction if tidal traffic flow schemes were adopted the same provision was made for all parts of the tunnel.

Experience had shown in the older tunnel that the length of relatively still air midway between ventilation stations could cause unpleasant atmospheric conditions. Accordingly, provision has been made for a supply of 80,000 cu feet/min (2,265 cubic metres/min) in addition to the normal 200 cu feet/min per feet (5.66 cubic metres/min) to be delivered midway along the tunnel. The maximum supply at each of the two stations is 740,000 cu feet/min (20,950 cubic metres/min) and the exhaust fans have a capacity about 10% greater.

The energy required varies as the cube of the velocity, and therefore of the quantity, unnecessarily high volumes of air are to be avoided. The fans have accordingly been designed for four-speed operation providing supplies in the ration 4:3:2:1 accordingly to traffic requirements. A pair of fans, of vertical spindle axial flow type, is installed at each of the shafts, mounted on a traversing carriage so that 100% standby capacity is available and either fan may readily be move into the duty position.

The details are as follows:
- Supply fans Exhaust fans
- Diameter205 inch240 inch
- Motor h.p450150
- Maximum duty pressure water gauge2.7 in.0.8 in.
- No. of fans (two stations) (one tunnel)44
- Supply fans metric conversion as follows 205 inches (5.2 metres) and 240 inches (6 metres)

The cut-and-over section at the Liverpool portal is separately ventilated. Fresh air enters naturally from the open cut approach and foul air is exhausted at roof level by eight jet fans spaced out over the 800 feet (243.8 metres) length, each of a capacity of 50 000 cu feet/min (1,416 cubic metres/min) and powered by a 6 hp motor.

Under the ventilation stations on the Seacombe and Liverpool shores, there are shafts and complex passage arrangements. At the junctions of the air supply ducts with the tunnel the segments invert gives place to lengths of in situ reinforced concrete construction with flat floor slabs between vertical sidewalls, forming the so-called square invert sections.

The TBM drive was a process operation and it was most undesirable to interrupt its progress by the special work required for the square invert sections, so these and the associated horizontal shaft and connecting passages had to be constructed either ahead of the drive or after the TBM has passed. The permanent work at the two square invert sites is similar, but at Liverpool the plan allowed little time after the passage of the TBM and before the construction of the road deck, so that as much of the work as possible had to be completed before the TBM arrived. This site required the more complex and interesting method of construction.

The pilot tunnels was used to allow access to the area for the working shaft and from the two ventilation shafts after these were sunk to the required level. The general idea behind the scheme was to construct the vertical reinforced concrete side walls of each section of square invert in a series of headings opened out of horizontal shaft driven from the shafts. The flat invert would be constructed by excavating down the pilot tunnel between the completed vertical walls.

One consequence of the scheme required a re-design of the permanent work, along with changes of access to allow necessary work to proceed in the shafts. Additional problems were encountered with the rock showing faults throughout. Only very restricted explosive charges could be used in shaft sinking given the restricted working conditions and the possibility of the faults widening and causing additional problems.

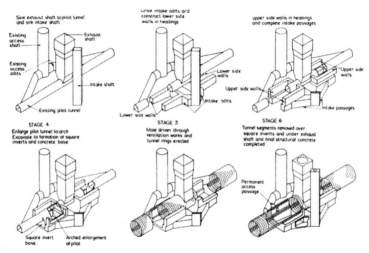

Liverpool Ventilation Construction (McKenzie Dodd)

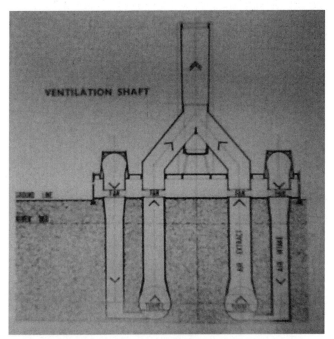
Liverpool Ventilation Shaft (McKenzie Dodd)

Liverpool Vent station (Author)

Wallasey vent Station (Author)

Chapter 15
Mersey Tunnel Police

For the first few years of its existence, the Mersey Tunnel Police followed the tradition of the Home Office police forces with high neck tunic jacket displaying the officers force collar number in chrome. The only difference in the uniform (due to the motorcycle combinations) was that motorcycle officers wore jodhpurs with detachable leggings. The leggings extended from the ankle to the top of the knee and were held in place with black metallic fasteners at the back. When the motorcycle combinations were phased out the wearing of jodhpurs ceased, since that time the uniform has favoured the Liverpool City Police, today known as the Merseyside Police.

In the 1960s, a white top to the cap was introduced this was detachable for easy cleaning and made of cotton, later on a one-piece cap with a white plastic top was introduced and has remained in service to the present day. In 1991, the diced black and white cap band was adopted.

With the introduction of the open neck, the tunic jacket the officers force number was positioned above the Mersey Tunnels Joint Committee on the epaulettes. In 1996, Paula Darlington became the first female civilian brought in to manage the Mersey Tunnels Police and through her efforts the addition of the Queens Crown came into being on the cap badge, she was also responsible for obtaining the police duty belt, speed cuffs and expandable baton etc. The general public could be forgiven in not noticing any difference in dress and personal equipment to that of their regular police colleagues.

When the Birkenhead (Queensway) was opened the police force was in effect a civilian force who used motorcycles and carried no emergency equipment. This may seem a little strange but given the vehicles of the time and small number using the tunnel, this was an appropriate response. In the 1940s there was a move to Ford vans, which were black in colour and had windows either side the length of the van which had police stop signs on top front and rear that could be illuminated. The vehicles carried basic emergency equipment, towrope, and bucket of sand, shovel, axe and two fire extinguishers. It was not until 1955 that Land Rovers were first brought into service. These were specially adapted with the back of the cab cut away which gave all round better vision for the driver, sides that dropped down and a top that lifted up for easy access to emergency equipment. These vehicles carried the same level of equipment as the Ford vans but with the addition of extra fire extinguishers and a 'skate' (this was a device made a steel with four small wheels that was used if a car or van had a flat tyre). The colours of these vehicles were cream with brown front wing panels that displayed the Mersey Tunnels Joint Committee crest on either sides also each door had the wording Mersey Tunnel Joint Committee, by 1965 the front panels had been sprayed cream.

This type of vehicle was unique to the Mersey Tunnel Police and was to remain in service with the addition of stream lining until the mid-1970s when it was replaced with the standard Land Rover that could carry more emergency equipment. These vehicles were used in both tunnels. The Land Rover was to remain in service until 1991 when the Land Rover Discovery, which is still in service today, together with Ford Transit vans, the workhorse of the Mersey Tunnel Police, along with the Ford Galaxy.

The Mersey Tunnel Police of today are a far cry from original tunnel police in 1936, as they now number 51 sworn officers in total, 1 Chief Superintendent, 1 Chief inspector, 5 Inspectors, 15 Sergeants, with the remainder made up of male and female Constables. The police are divided into five teams or sections working a three-shift system, giving 24-hour cover under the supervision of a sergeant or inspector. The senior officers are always on call should if the need arises. The responsibilities of the tunnel police include policing, breakdown services and toll collection as well as a multitude of disciplines people would not normally associate with a tunnel.

All Mersey Tunnel Police vehicles carry quite a comprehensive range of emergency equipment but the Ford Transit's are used to escort dangerous loads through the tunnels. All livery and markings are the same as the Home Office police forces throughout the U.K. The vehicles currently in use (June 2014) are nine and are as follows. Ford SMax x 5, Land Rover Discovery x2, Ford galaxy x 1, and a Mercedes Vito x 1 which is used for prisoner transportation.

Mersey Tunnel Police have statutory power to appoint members of the tunnel staff as special constables. *Over 100 employees with police powers are organised as a small police and traffic control unit with a superintendent, ten inspectors, and nine sergeants.* With 50,000–60,000 daily journeys, the tunnel police ensure safe and smooth passage by enforcing tunnel by laws and special tunnel operating procedures.

Inside the tunnel, the police patrol in vehicles equipped to deal with breakdowns, fires and other emergencies. These patrol vehicles are equipped with short wave radio and are controlled by the inspector on duty, who has television surveillance inside the tunnels and over the approach roads.

The patrols normally deal with:
- Any emergencies, within the tunnel or approaches on both sides of the river
- Direct traffic and enforce the regulations, including maximum and minimum speed limits
- Prohibition of lane changing within the tunnel.
- Police the entrances prevent the entry of pedestrians and pedal cyclists.
- Enforce the regulations regarding the passage of dangerous goods and out of gauge (wide) loads
- Check the loading and mechanical condition of vehicles.
- In addition, they make a contribution to general law and order in the vicinity of the tunnels,
- Apprehending drivers of stolen vehicles carrying stolen goods,
- Dealing with drunken and disorderly persons etc.

Over 1000 cases are taken to court for offences in the tunnel or surrounding road system operated by the Mersey Tunnel Police. In the Queensway Tunnel, which has two lanes in each direction with solid white lines separating the lanes, lane changing is prohibited to reduce the risk of accidents and to maximise throughout.

The Wallasey tunnel has a two-lane system in each tunnel, again a solid white line separates the lanes, and there is a strict no overtaking rule in both tunnels. This system is changed during rush hour to three lanes in one direction and one lane in the opposite direction. During morning peak the majority of lanes is in the direction of Liverpool and evening peak Wirral. The appropriate amount of traffic cones and tunnel police direct the traffic to the correct lanes system in operation at the time.

This is enforced by a double white line down the centre of the carriageway. Under normal operations, traffic will use one tube for each direction of travel but it is expected that the 'Keep in Lane' rule as per the Queensway and double white lines will be retained. If overtaking is ever permitted, it would probably prevent the flow of traffic as this would be highly dangerous if tunnel users become accustomed to changing lanes. A strict speed limit is also enforced in both tunnels (Queensway 30mph and Kingsway 40mph) and the speed limit restrictions also apply to the approach road, which are appropriately signed and regulated.

In the early year's responsibility for policing of the Birkenhead Queensway Tunnel was carried out on a 6-month basis by both then Liverpool City Police and Birkenhead Borough Police. This created a number of problems with the jurisdiction and the manner in which the forces authority differed. The boundary of the Mersey Tunnel fell between two counties, Liverpool in the county of Lancashire and Birkenhead in the county of Cheshire (the county boundary being near the centre of the tunnel). The duty police officers were reluctant to take directions from the Mersey Tunnel inspectorate, as they were not police officers so this caused operational problems.

Neither side was happy with the arrangements so the Mersey Tunnel Authority asked the Home Office for permission to form their own police force. This request was granted and the Home Office under Section 105 of the 1936 Liverpool Corporation Act. The Home Office made it clear the cost of a tunnel police force would be borne by the MTJC and not Central Government. The Act states the Mersey Tunnels Joint Committee shall have the power to appoint any of their officers or servants to act as special constables for the policing of the tunnel, its approaches and areas governed by the Mersey Tunnel by-laws.

This act caused yet more problems as Special Constables cannot receive payment as a traffic officer as described in the Mersey Tunnel by-laws, but held the office of constable. To comply with the jurisdiction of policing both counties the Mersey Tunnel Police Officer had to swear an oath in both Liverpool and Birkenhead. In January 1936, the Mersey Tunnel Police came into being taking over from Liverpool City Police and Birkenhead Borough Police. The Mersey Tunnel Police initially had 2 Inspectors, 4 Sergeants and 14 Constables who carried out regular patrols using motorcycle combinations.

A 'running card' system which detailed the patrol to take place at specific times of the day from each end of the tunnel. In theory at least two patrols would be inside the tunnel travelling in opposite directions at regular intervals (approximately every 15 minutes). Patrols had to use the side lane which was then classed as the 'slow lane' as heavy goods vehicles had to use this lane. The patrol would usually be out of the tunnel in 4 minutes, as a consequence of this a patrol was never at the right entrance at the right time. All Mersey Tunnel Police officers gave a sigh of relief when in 1972 the running cards were phased out, trusting a police officer to carry out patrol under their own initiative.

Mersey Tunnel Police ensure that traffic using the tunnel or approach roads do so as quickly and safely as possible and to deal with anything or persons that prevent it from doing so. Before the introduction of CCTV a breakdown if not dealt with quickly could lead to a backlog of traffic. This leads to frustrated drivers changing lanes and trying to push their way through the solid traffic at the tollbooths.

The tunnel police, as well as mobile patrols supervise all traffic entering to ensure compliance with the Mersey Tunnel by laws. As would be the case in most tunnels and confined spaces, explosive, flammable and other such materials are closely controlled. The Mersey Tunnel Police at the entrances are ensuring that drivers do not attempt to use the tunnel through ignorance or blind ignorance of the byelaws to avoid travelling to the Runcorn Bridge route which is much longer.

During the morning peak, 8.25am–9.15am, a system of allowing traffic in one direction to flow easier is in use. This entails 50% of the carriageway in the opposite direction given up the major flow of traffic this is more the case of the Wallasey (Kingsway Tunnel) where three lanes of traffic are allowed to enter the tunnel from Wallasey as people made their way into work in Liverpool.

In the 1950s this was also the case for the Birkenhead tunnel (as it was the only one at the time) and the filtering of traffic was done using large black boards with a large white arrow pointing to the next lane (behind these boards putting on a brave face an officer wearing a white coat).

The boards were strategically placed in the centre of the lane indicating the vehicles approaching to filter into the next lane. If the flow of traffic was a steady flow it would be moving at approximately 15 to 20 miles per hour not fast under normal circumstances but terrifying fast in the confines of a tunnel. You have to admire the Mersey Tunnel Police Officer at the time as he waved traffic into the next lane praying that all the drivers were paying attention to the directions. It has to be said that it was not rare for the boards to be clipped and it was more good luck than good management that no officers were ever killed or injured. As if this was not enough, they also had to cope with exhaust fumes from the vehicles in and around the tunnel.

Any breakdown initiated the calling of a recovery vehicle, which, on occasions, would take up to an hour to reach the broken-down vehicle. This would cause the traffic to build behind the breakdown in an already congested tunnel. If the breakdown occurred whilst the single lane was in operation grid locked would quickly build up on the relevant side of the river depending on the time of day.

In the early years of the tunnels, the Mersey Tunnel Police would walk to the vehicle and this would include breathing in all the fumes and bad air in the tunnel. They would also be breathing in this foul air during the duration of the breakdown until the vehicle has been recovered.

During the mid-60s, CCTV cameras were installed in the Birkenhead (Queensway) Tunnel to assist the Police in their duties and dealing with day to day occurrences and incidents. This was accomplished by the Police in the new Police HQ ad control room at the tunnel. This was to be the case until 1971 when the Wallasey (Kingsway Tunnel) opened with a purpose-built control and admin section above the tollbooths.

In the early years, Mersey Tunnel Police were thought not to need much (if any) training so this was taken as more of a hands-on approach especially given the low levels of slow moving traffic compared to today. Today, the Mersey Tunnel Police have to conform to a large number of statutory traffic and other regulations similar to the police you see on the street. In addition to this, they also cover the regulations associated with the tunnels. With today's Mersey Tunnel Police, the initial training course is sixteen-weeks long and includes the normal practice of shadowing (as a probationary officer) an experienced officer. The course is in modular format and follows very closely Home Office training. Because the Mersey Tunnels Police is a small force compared with others, there was never a need to sit promotion exams. Senior officers are well aware of suitability of an officer to fill the vacancy. Tunnel

Police Officers would be expected to attend a selection board before a decision was made.

The Mersey Tunnel Police also train their officers to deal with any incident or situation that may arise in such a confined space as a tunnel. One of the greatest and most feared of incidents in any confined space would be a fire or explosion. This can be shown in the devastating effect from the Montblanc and Euro tunnel Fires. Both spread quickly and with devastating effect to both vehicles and lives.

Mersey Tunnel Police would be the first on the scene and are trained in both firefighting and first aid. These skills are important part of their initially training programme with refresher courses being held every couple of years. Training is constantly being updated. Mersey Tunnel Police can claim the distinction of being the best trained and equipped non-Home Office police service in the U.K.

Early 2015 brought in a new milestone for the Tunnel Police with the opening of a new state of the art control room. The new control room has a state of the art Video Wall which is fed by over 400 cameras which ensure you are never alone in the tunnels. There are 800 cameras covering the areas over the whole network for which the Mersey Tunnels are responsible. The tunnel police vehicles are also state of the art and packed with such things as safety and medical equipment, including defibrillators.

Despite the requirement for no overtaking where two solid white lines within the Highway Code the tunnels have this policy in both tunnels for safety reasons.

Some of the unknown incidents the tunnel police have to deal with are:
- People suffering from Claustrophobia within the tunnels
- People walking through a road tunnel
- Cycling through a road tunnel

Cycling is allowed through the Queensway (Birkenhead) Tunnel between 8 p.m. and 8 a.m. on a Saturday night and other nighttime hours through the week, but cycling is not allowed in the Kingsway (Wallasey) tunnel.

There was one particular incident in late 2015 when four youths used City bikes to try and get home on 27 December after a night out in Liverpool. One of the youths later stated that it was a bit stupid and dangerous and it would not happen if they were sober. It was his brothers leaving the party before he went to Australia. They were unable to get a taxi so decided to use the bikes to get home and return them in the morning.

There are a number of incidents each year of people looking to 'walk home' or 'walk to work' and they are unable to get some form of transport at that time of the day for a variety of reasons. In 2015, there were over 153 'pedestrian incidents' as people paid with not only their safety, but the safety of other tunnel users. Of this number, 82 were in the Queensway (Birkenhead) Tunnel and 71 were in the Kingsway (Wallasey) Tunnel, all between April and October.

Mersey Tunnel police will quickly see the problem on the numerous CCTV and quickly apprehend anyone trying to enter the tunnel and an alarm system is also in place to alert them to such incidents. The Police will guide the person to a safe area in one of their patrol vehicles and ensure the tunnels are performing safely and efficiently at all times. Between 2011 and 2014, there were 300 such incidents in the Queensway (Birkenhead) Tunnel.

Chapter 16
Tolls Finances Maintenance and Repairs

The tunnel tolls are there to prepay the original debt along with the numerous other costs that have been incurred through the years. The Integrated Tunnel Authority is responsible for determining the levels of tolls.

As you drive to the toll booth, you will see the name of the booth operator on a display at the bottom of the window. It is always nice to say hello and give the person a name and pass the time of day. It is not this person who is responsible for what is seen as a tax on the motorist, but the committee who set the tolls on an annual basis. The booth operator is only doing a job at the end of the day no matter what you think of the tolls and the ever-increasing costs. The money you place in a bucket is sent via tubes to a central money room. The cash is then counted and bagged for transportation to the financial institutions.

The tunnels were making losses for a multitude of reasons and such as the cost of the new tunnel, a recession and thus less traffic as people started to tighten their belts. In 1972, an Act was passed for limited powers were added so the additional losses could be added to the existing tunnel debt, but these powers were limited and ran out in 1980. In 1975, tunnel tolls were increased to 20p, for cars and 1975 to 25p and 1979 to 30p.

The year 1980 saw an Act being passed which consolidated various old Acts and the opportunity was taken to add losses to debt. Tolls were raised again in 1981 by 40p and again in 1986 to 50p. Despite the toll increases, the debt increased and this was primarily due to the high inflation rates at the time. The tolls are still seen as a burden on the local economy, but they are more of a necessity given the easy access to both sides of the river. In 1999, the tolls increased to £1.20. The current toll (2016) stands at £1.70 for cars.

In 1975, the tunnels levels of use were back down to 1959 levels and it was thought that this decline was part of a rolling cycle of world events including the oil crisis of 1972–73. The tunnels use fell again in 1984, as they were now less than 25 million vehicles a year.

In the md 1970s, it was decided that the buses would be able to use the tunnels and this was followed by de-regularisation on 1986, which saw more and more buses using the tunnels. By November 1978, there was shock across Merseyside when it was announced that there was consideration being given to the tunnels closing to prevent the authority from going into bankruptcy such was the fine financial balancing act at the time.

In July 1968, hopes of an early end to the Mersey Tunnel tolls was all but killed off by the Government. The cabinet decided to reject a plea to lift the tolls in an influential all party committee of MPs. The department of Transport was in favour of phasing out the charges but the treasury was keen to block any changes. The main argument was that spiralling debts at Britain's 10 major tunnels and bridges could not be seen as a priority with the government trying to keep a firm grip on public spending.

The government had put the total cost at £750 million but the MPs put the figure at £356 million. At this, the government decided to cut tolls on the Humber Bridge and Erskine Bridge across the Clyde on the grounds that the debt could never be paid off. The difference in the costs was not explained and why the two selected bridges were given priority over the Mersey Tunnels was somewhat strange.

Echo Article from April 12 1944:

'The Mersey Tunnel is a national asset which has proved its value in peace and war. Its existence is due to the initiative and enterprise of the citizens of Merseyside, who are shouldering heavy liabilities as a result of such enterprise. Tolls are imposed to meet loan charges, but such an arrangement is not in the public interest, and Birkenhead Council favours the promotion of legislation to abolish these tolls. Alderman Sherman, as chairman of the Mersey Tunnel Joint Committee, like every other leader on Merseyside, would like to see the Tunnel tolls abolished, but before this can be done the financial responsibility for the undertaking would have to be transferred from local rates to national funds.

The time is approaching when the Mersey Tunnel should be nationalised, but the Chancellor of the Exchequer and the House of Commons have to be converted to this view. It will be necessary to press the case with vigour at the appropriate time, otherwise the ratepayers of Merseyside will be left to carry the burden. Liverpool is already one of the most heavily rated cities in the country, largely owing to circumstances beyond our control, and nobody knows what effect the new Education Bill and other measures of social reform will have on our finances.

Merseyside suffered severely between the two wars because the maintenance of the able-bodied unemployed was made a local and not a national charge. In the past, the heaviest burdens have been heaped on those least able to bear them. If the Mersey Tunnel is a national asset—and nobody has come forward to deny that claim—the financial responsibility should be borne by the National Exchequer. The whole problem of rates and local government is in the melting pot and Merseyside is in a position to make notable contributions to the shaping of future policy.

We have been beggared in some respects but we have a wealth of experience on which to draw. Liverpool gave a lead to the rest of the country in laying down the policy that the maintenance of the able-bodied unemployed should be a national charge and we do not accept dictation from Whitehall. Merseyside is always willing to pay its share and desires a square deal, but experience has taught us that this does not descend like the gentle rain from heaven. Plain speaking, bold planning, and vigorous action are necessary if we are to obtain our just dues both as regards the Tunnel and other enterprises of national importance.'

The County of Merseyside act, which was amended by the 2004 Mersey Tunnels Act sets out the procedure for revising the tolls. Under Section 91(7) of the 1980 Act, the authority must make an order in February of each year (a Section 91 Order) to fix the amount of tolls payable.

Within this order, they must specify which of the four classes is to pay what amount of money to use the tolls.

CLASS	SUB CATEGORY	
1	(a)	Motor Cycle with side car and three-wheeled vehicle
	(b)	Motor Car and goods vehicle up to 3.5 tonnes gross weight
	(c)	Passenger vehicle other than a motor car with seating capacity for under 9 persons
2	(a)	Motor car and goods vehicle up to 3.5 tonnes gross weight with trailer
	(b)	Goods weight over 3.5 tonnes gross weight with two axles
	(c)	Passenger vehicle with seating capacity for 9 or more persons with two axles
3	(a)	Goods vehicle over 3.5 tonnes gross weight with three axles
	(b)	Passenger vehicles with seating capacity for 9 or more persons with three axles
4	Goods vehicle over 3.5 tonnes gross weight with 4 or more axles	

The general consensus on the tolls is that they are a further road tax for those who use the tunnels and as such stifle the local economy. They are also seen as allowing Mersey Tunnels to charge what they like and use the money as they see fit. The general presumption amongst others is that the 1980 Act (as amended) is that the tolls rise in line with inflation, and therefore preserve the value in real terms. The increase in tolls is authorised by the base toll amount as defined in section 91 (6) of the 1980 Act. It also allows for the same increase as the RPI (Retail Price Index).

Every time the authority makes a section 91 order, it must also consider a section 91C. This section 91C gives the discretion to reduce the amount of the tolls payable by any particular class of vehicle using the tunnel. The reasons for this reduction may vary and is defined as necessary or appropriate and have regard to economic or social nature to the County of Merseyside.

When deciding on the amount of tolls to charge, the authority would have to keep within the scheme and purpose of the scheme originally founded upon by the authority and accepted by Parliament. The cost of the tunnels keep pace with other transport alternatives, but at varying stages through time the members as well as the public have not believed this is the case.

The Toll Booths are now situated at the Birkenhead side of the 3.24 kilometre (2.01 miles) tunnel, but originally they were situated on both sides of the tunnel.

Class	Toll	Nov 99	Apr 05	Apr 06	Apr 07	Apr 08
1	Authorised	£1.20	£1.40	£1.40	£1.40	£1.50
	Cash	£1.20	£1.30	£1.30	£1.30	£1.40
	Fast Tag	£1.10	£1.15	£1.15	£1.15	£1.15
2	Authorised	£2.40	£2.70	£2.80	£2.90	£3.00
	Cash	£2.40	£1.30	£1.30	£1.30	£2.80
	Fast Tag	£2.20	£1.15	£1.15	£1.15	£2.50
3	Authorised	£3.60	£4.10	£4.20	£4.30	£4.50
	Cash	£3.60	£3.90	£3.90	£3.90	£4.20
	Fast Tag	£3.30	£3.15	£3.45	£3.45	£3.75
4	Authorised	£4.80	£5.40	£5.60	£5.80	£6.00
	Cash	£4.80	£3.90	£5.20	£5.20 3.20	£5.60
	Fast Tag	£4.40	£3.45	£4.60	£4.60	£5.00

Apr 09	Apr 10	Apr 11	Apr 12	Apr 13	Apr 14
£1.60	£1.60	£1.60	£1.70		
£1.40	£1.40	£1.50	£1.50	£1.60	£1.70
£1.25	£1.25	£1.30	£1.30	£1.40	£1.40
£3.10	£3.10	£3.30	£3.40		
£2.80	£2.80	£3.00	£3.00	£3.20	£3.40
£2.50	£2.50	£2.60	£2.60	£2.80	£2.80
£4.70	£4.70	£4.90	£5.20		
£4.20	£4.20	£4.50	£4.50	£4.80	£5.10
£3.75	£3.75	£3.90	£3.90	£4.20	£4.20
£6.20	£6.20	£6.50	£6.90		
£5.60	£5.60	£6.00	£6.00	£6.40	£6.80
£5.00	£5.00	£5.20	£5.20	£5.60	£5.60

Solo motorcycles toll free
Vehicle Classes and Tolls

Commencing 00.01 hours, Monday 1st April 2013, until further notice:

Toll Class	Description	Cash Toll	Fast Tag Toll
1	Motorcycle with sidecar and 3-wheeled vehicle Private/light goods vehicle up to 3.5 tonnes gross vehicle weight Passenger carrying vehicle with seating capacity for under 9 persons	£1.60	£1.30
2	Private/light goods vehicle up to 3.5 tonnes gross vehicle weight, with trailer Heavy goods vehicle over 3.5 tonnes gross vehicle weight, with two axles Passenger carrying vehicle with seating capacity for 9 or more persons, with two axles	£3.20	£2.60
3	Heavy goods vehicle over 3.5 tonnes gross vehicle weight, with three axles Passenger carrying vehicle with seating capacity for 9 or more persons, with three axles	£4.80	£3.90
4	Heavy goods vehicle over 3.5 tonnes gross vehicle weight, with four or more axles	£6.40	£5.20

Vehicle Classes and Tolls

Commencing 00.01 hours, Sunday 06th April 2014, until further notice:

Toll Class	Description	Cash Toll	Fast Tag Toll
1	Motorcycle with sidecar and 3-wheeled vehicle Private/light goods vehicle up to 3.5 tonnes gross vehicle weight Passenger carrying vehicle with seating capacity for under 9 persons	£1.70	£1.40
2	Private/light goods vehicle up to 3.5 tonnes gross vehicle weight, with trailer Heavy goods vehicle over 3.5 tonnes gross vehicle weight, with two axles Passenger carrying vehicle with seating capacity for 9 or more persons, with two axles	£3.40	£2.80
3	Heavy goods vehicle over 3.5 tonnes gross vehicle weight, with three axles Passenger carrying vehicle with seating capacity for 9 or more persons, with three axles	£5.10	£4.20
4	Heavy goods vehicle over 3.5 tonnes gross vehicle weight, with four or more axles	£6.80	£5.60

Solo motorcycles toll free
Fees and Charges

Commencing 00.01 hours Thursday, 1 February 2007, until further notice:

Toll Class		Breakdown Charge
1	Motorcycle with sidecar and 3-wheeled vehicle Private/light goods vehicle up to 3.5 tonnes gross vehicle weight Passenger carrying vehicle with seating capacity for under 9 persons	£72 per hour Minimum charge one hour Penalty Surcharge £51
2	Private/light goods vehicle up to 3.5 tonnes gross vehicle weight, with trailer Heavy goods vehicle over 3.5 tonnes gross vehicle weight, with two axles Passenger carrying vehicle with seating capacity for 9 or more persons, with two axles	£104 per hour Minimum charge one hour Penalty Surcharge £77
3	Heavy goods vehicle over 3.5 tonnes gross vehicle weight, with three axles Passenger carrying vehicle with seating capacity for 9 or more persons, with three axles	£127 per hour Minimum charge one hour Penalty Surcharge £101
4	Heavy goods vehicle over 3.5 tonnes gross vehicle weight, with four or more axles	£145 per hour Minimum charge one hour Penalty Surcharge £109

For stoppages caused by fuel shortage or flat tyre (with no serviceable spare) there is a penalty surcharge in addition to the normal breakdown charge.

- Escort Fees
- Control of one lane: £15
- Control of two lanes: £39

On 26th July of 1976, the tollbooths became automatic in that a vast majority of them were now unmanned as an experiment by the then Merseyside County Council to save money.

At first, only cars were able to use the automatic booths with a cash only (no change given) policy in order. When a driver approached the booth the amount to be displayed was shown and this was to be dispensed in the basket, as is still the case today. At that time only genuine coins of the realm were accepted but there were a few people who tried (and some successfully) to defeat the system. An exception was made for the 1/2p and 2 1/2p (the old sixpence piece) pieces which were not accepted.

If the full amount was not paid in or spurious coins (items of metal such as washers etc.) were deposited the amount payable would still show on the indicator and the spurious or other items would be rejected into the basket below for the driver to remove. Once the amount was paid, a traffic signal changed to green and the barrier lifted to allow progress through the tunnel.

This system of tunnel tolls has had an additional automatic layer added which allows the drive to proceed through the tunnel via the automatic tollbooths. The system is called a 'Fast Tag' system, was introduced in 1993 and had previously seen similar systems successfully used in Norway, Texas.

In 2010, there was a report by Colin Buchanan on 'Mersey Tunnels Tolls – Evaluating the Impacts' ('the report'). The study focused on the socio-economic impacts of the Tunnels tolls, by profiling the current socio-economic conditions and Tunnels usage, and imputing the relationship between the two.

The report was an assessment of Merseyside's economic link between the trips to Liverpool City Centre from different boroughs on Merseyside. Retail analysis, including an assessment of the distribution of retail spend across Merseyside, to understand how this is influenced by the tolls. And a business survey, to learn how businesses perceive their location and tolls in relation to their business.

Volumes of traffic using the Tunnels have grown steadily over the decades, reaching a peak in 2005 and a small decline since, which mirrors the trend on urban roads nationally and the recent economic downturn. Congestion remains a problem, with 6 out of 10 Tunnels' users queuing during the morning and evening rush hours.

In April 2013, it was announced that a new tolling system could be introduced to collect fees from Tunnel users. The Transport Authority (Merseytravel) commissioned a study from Parsons Brinkerhoff at a cost of £51,700 as to the use of 'Eye in the Sky' toll collections via ANPR (Automatic Number Plate Recognition System) systems and the redevelopment of the Mersey Tunnels toll plazas and the possible introduction of new technology. Merseytravel stated that the current toll systems are nearing the end of their lifespan so the authority has decided to look into options for the Kingsway and Queensway Tunnels. The study looked into the redevelopment of the plazas and types of tolling systems and payment methods.

The existing tolls system was installed in both Queensway and Kingsway in 1990s with a major update of Queensway Plaza equipment in 2000 and Kingsway in 2002. The server network was upgrade in 2007 and was contracted for a 5-year period so this would see the planned upgrade as a normal consequence of such a complicated integrated system. The study looked at the plazas and types of tolling systems to be used which could benefit the users as an overall saving to the day-to-day tunnel costs and bring them into the 21st century. However, critics of the scheme stated that the present collection methods were working and questioned why this expensive system was being considered

According to the Merseyside Integrated Transport Authority / Executive (Savings and Development 2014-15) the following would apply as far as proposals for savings within the overall group. I have only concentrated on the tunnels and this is relevant to this publication. I have also explained the breakdown of other areas, as they are relevant to the overall picture. There are discussions as to updating the existing system and use a smartcard system, which will include an ANPR which will be linked the DVLA (Driver and Vehicle Licencing Authority).

This system was introduced (November 30th 2014) on the Dartford Crossing in London.

Service Growth / Savings

	2014/15 Revenue Costs (£000)	Ultimate Annual Revenue Costs (£000)
SUMMARY		
Democratic Representation and Corporate Management	(70)	(70)
Network Management Bus Services	(1,722)	(1,722)
Network Management Rail Services	(5)	(5)
Travel Concessions	(256)	(256)
Customer Service Hubs	(150)	(150)
Mersey Ferries	(167)	(167)
Mersey Tunnels	(2,657)	(2,657)
Asset Management	–	–
Policy and LTP (Local Transport Policy) Development	(8)	(8)
People & Customer Development	(146)	(146)
Resource Directorate		
Funds Management	2,223	2,223
TOTAL	(2,958)	(2,958)
Public Transport Services		
Service Savings / Income Growth	(2,958)	(2,958)
Service Growth / Development		
TOTAL	(2,958)	(2,958)

The specific area of the savings relating to the tunnels is as follows:

	2014/15 Revenue Costs		Ultimate Annual Revenue Costs	
	£000	£000	£000	£000
Savings in Casual Overtime Costs		(30)		(30)
Savings in Publicity and Promotion	(5)		(5)	
Savings in AVI Fast Tag Issues	(40)		(40)	
Savings in AVI Mailing / Postage Services	(22)		(22)	
Savings in Disposal Services	(10)	(77)	(10)	(77)
Savings recharged From the Following				
Asset Management		(128)		(128)
People and Customer Development		(18)		(18)
Resources Directorate		(9)		(9)
Tolls Increase		(2,395)		(2,395)
TOTAL		(2,657)		(2,657)

It would follow that the 2013/14 revenue budget, along with the 2014/15 revenue budgets would be as follows according to the Merseyside Integrated Transport Authority / Passenger Transport Executive Revenue Budget:

2013/14		
Approved Budget	Restated Budget	Revised Estimate
£000	£000	£000
3,683	3,683	3,667
786	786	804
155	155	148
980	980	1,037
271	271	275
90	90	95
25	25	50
563	563	629
543	543	576
289	289	290
151	151	151
5,542	5,542	4,646

13,078	**13,078**	**12,368**
374	374	396
678	678	718
1,255	1,255	563
		437
37	37	48
15,422	**15,422**	**14,530**
5,291	5,291	5,528
3,663	3,663	3,663
9,000	9,000	7,875
18,584	**18,584**	**17,066**
34,006	**34,006**	**31,596**
38,750	38,750	38,995
440	440	327
39,190	**39,190**	**39,322**
(5,184)	**(5,184)**	**(7,726)**

Service Details	2014/15		
	Base Estimate	Growth (Savings)	Drafts Budget
	£000	£000	000
Operating Expenditure			
Employees	3,583	(30)	3,553
Salaries and Wages	799		799
National Insurance etc			
Premises			
Repair and Maintenance of Buildings and Grounds	148		148
Power	1,138		1,138
Rates, Cleaning etc	266		266
Premises Insurance	95		95
Supplies and Services			
Equipment	50		50
Printing, Telephones etc	645	(77)	568
Operational Tools and Equipment	536		536

Plant and Machinery Insurance	290		290
Transport & Plant	150		150
Asset Man'gt, Repairs and Project recharge	5,151	(128)	5,023
Provision for Restructure and other Contingencies			
Sub Total	**12,851**	**(235)**	**12,616**
Customer Services ICT	433		433
People and Customer Development	812	(18)	794
Resources Directorate	612	(9)	306
Customer Delivery Hubs	513		513
DRCM	49		49
Total Expenditure	**15,270**	**(262)**	**15,008**
Capital Financing			
Debt Charges	5,290		5,290
Levy Repayment General Fund	3,663		3,663
Revenue Contribution	8,000		8,000
Total Capital Financing	**16,953**		**16,953**
Total Expenditure	**32,223**	**(262)**	**31,961**
Operating Income			
Toll Income	38,995	2,395	41,390
Fees and Other Charges	327		327
Total Income	**39,32**	**2,395**	**41,717**

NET EXPENDITURE	(7,099)	(2,657)	(9,756)

Source Accounts/STD/MITA Budget 2014

Merseyside Integrated Transport Authority / Passenger Transport Executive, Revenue Budget

Analysis of toll Income and Expenditure 2013–2015

	Approved Budget 2031/14	Revised Budget 2013/14	Draft Budget 2014/15
	Nos 000	Nos 000	Nos 000
Analysis of Traffic Paying			
Class 1 up to 3.5 Tonnes	22,805	23,321	23,321
Class 2 over 3.5 Tonnes (2 Axles)	626	612	612
Class 3 over 3.5 Tonnes (3 Axles)	90	99	99
Class 4 over 3.5 Tonnes (4 Axles)	352	359	359
Sub Total	23,873	24,391	24,391
Concessionary	450	448	448
Traffic Total	**24,323**	**24,839**	**24,839**
	£000	£000	£000
Analysis of Toll Income			
Class 1	34,745	34,970	37,069
Class 2	1,710	1,668	1,791
Class 3	375	408	438
Class 4	1,920	1,949	2,092
Total Income	**38,750**	**38,995**	**41,390**
	Nos	Nos	Nos
Staffing			
Traffic Control and Law Enforcement	52	52	51
Toll Collection and Administration	64	64	60
Total	**116**	**116**	**111**

Source Accounts/STD/MITA Budget 2014

According to the Merseyside Integrated Travel Authority Accounts of 31 March 2014, the following statements were made for the overall Authority.

Key Performance Indicators and Business Review

The financial results of the Authority and group (comprising of the Authority, and the Merseyside Passenger Transport Executive and subsidiaries) accounts for budgeting purposes, is based upon the income and expenditure of the general reserve and shows a surplus for the year £0K (2012/12 £0K). This basic differs from that shown in the financial statements. The MITA (Merseyside Integrated Transport Authority) group position shows a surplus of 376k (2012/13 deficit £7k). The Un/Usable reserves of the Authority at the yearend were £115m (2012/13 £80m). The Group reserves were £208m (2012/13 £163m).

Service Provision

During the year, net expenditure of £127.4m (funded by the Levy – local taxation) was incurred for the provision of transport services consisting of rail, ferries, tunnels, supported bus services and concessionary travel arrangements for elderly, disabled and children. With the exception of the Authorities direct services (the Mersey Tunnels and servicing of £0.2bn Tunnels and Transport Infrastructure Debts), all other transport services are secure through the Executive which is financed by Authority Grants.

Cash flow

Although, base rate continues to be at a historical low levels, interest of £1.1m was earned during 2013/14 (£1.4m 2012/13) through prudent short-term deposits of surplus monies. The ITA (Integrated Transport Authority) continues to benefit from the surplus PTE monies being on-lent to the ITA interest free, to permit bulk placements on the money market thereby maximising investment opportunities and returns for Merseytravel.

Mersey Tunnels

The Mersey Tunnels Act 2004 permits any operating surplus to be utilised by the authority to achieve public transport policies in its local transport plan. In 2011/12, £2.5m was transferred into the Authorities General Fund (2010/11 £5.3m) and a tunnel reserve and renewals fund balance of £7.3m remained as at 31 March 2012, (£5.7m 31/3/11).

Prior to the tunnel operations becoming self-financing, the tunnels deficit was funded by the Authority Levy. Over several years levy funding reached £28m, which following legal advice is now being repaid by Tunnels over 21 years. The Authority has approved an annual repayment, including interest of £3.6m.

Investments

NAME	IMMEDIATE PARENT	GROUP HOLDING
Mersey Ferries Ltd	Merseyside Passenger Transport Executive	Company Limited by Guarantee
Mersey Passenger Transport Services Ltd	Merseyside Passenger Transport Executive	25 £1 Ord Shares and 375 £1.5% Non-Cumulative Pref. Shares
Real Time Information Group Ltd	Merseyside Passenger Transport Executive	1 £1 Ord Shares
Global Smart Media Ltd	Merseyside Passenger Transport Executive	17,648 – 10p Ord shares
Accrington Technologies Ltd	Merseyside Passenger Transport Executive	500 £1 Ord Shares
The Beatles Story Ltd	Mersey Ferries Ltd	290,000 £1 Ord Shares

NATURE OF THE BUSINESS	% EQUITY INTEREST 2014	% EQUITY INTEREST 2013
Passenger Transport	N/A	N/A
Leasing	100	100
Real Time Information Systems	100	100
Smartcard	87.9	87.9
Smartcard	50.1	50.1
Tourism	100	100

Source: Merseyside integrated Travel Authority Accounts 31 March 2014.

To allow the above and main business to run, the five main directors have the following remuneration costs attributed to them.

Post Title	Year	Salary £
Chief Executive / Director General	2014	127,309
	2013	77,367
Director of Resources	2014	92,154
	2013	
Director of Integrated Transport (& Deputy Chief Executive)	2014	126,462
	2013	77,367
Director of Corporate Development	2014	104,188
	2013	99,185
Director of Customer Services	2014	0
	2013	103,156

Allowances £	Pension Contribution £	Total Remuneration £
1,717	15,929	44,955
61,132	12,696	151,195
	11,519	103,673
		0
1,954	15,885	144,301
35,849	10,524	123,740
628	13,102	117,918
9,100	13,536	121,821
0	0	0
14,342	14,529	132,027

Source Merseyside integrated Travel Authority Accounts 31 March 2014

Mersey Tunnels Claims and Myths as Denoted on the Mersey Tunnels Web Site, November 2014

1 Claim: The tunnels debt was paid off long ago

Fact: The loans are still being paid and will not be paid off until 2048. These payments are fixed and the penalties incurred by paying them off early would make the debt more expensive and therefore be a false economy. The debt does not just relate to the construction costs of both the Queensway and Kingsway, but also the money that was borrowed for tunnel maintenance and improvements before the Tunnels Act 2004. This now ensures sufficient funds are generated through tunnel use to allow for their upkeep and improvement without borrowing.

As assets, the tunnels are worth around £1bn. The capital expenditure levels and maintenance being paid is modest considering their value.

2 Claim: There are no benefits for local users

Fact: The tunnels offer discounted journeys to those using our Fast Tag scheme. The more journeys people make through the tunnel the more users will save, so it's particularly beneficial for the most regular users, such as commuters – most likely to be local users.

The Fast Tag toll is almost one fifth cheaper than the cash toll, meaning for every six journeys, the equivalent of the seventh journey is free. More than a third of payments (35%) are already made by Fast Tag, meaning many tunnel users already benefit from a reduced toll rate each journey.

3 Claim: The tunnels could be toll free as part of the national road network

Fact: Only those road projects, including tunnels, built on approval from Government as part of the national road network are paid for indirectly via road tax and general taxation. The tunnels were built as a joint venture by local authorities and as such their upkeep and maintenance remains outside of the national motorway network. Calls have been made locally to Government to review this, but there are no plans currently.

4 Claim: Toll rises are out of line with inflation

Fact: Under the Tunnels Act 2004 tolls should rise by the level of inflation, but aren't permitted to go above this. The toll level increases since the Act was introduced in 2004 have been below the level of inflation as elected members have exercised their discretion on economic and social grounds to maintain a discount. For example, Mersey travel's integrated authority should have increased the toll up to £1.80 in 2013, but instead only approved an increase of 10p to £1.60. Being in line with inflation keeps the toll broadly in line with other cross-river travel costs such as bus and rail fares.

5 Claim: The tunnels haven't improved since they were opened

Fact: There are more than 25 million vehicles a year through the tunnels and significant investment in the operation, maintenance and development is needed to keep them safe and efficient. They are regarded as the safest tunnels in the UK and amongst the safest in Europe for their age following the last independent assessment. Major investment over the last seven years has involved the creation of new escape chambers for use in the event of an accident, resurfacing of the Queensway Tunnel, re-cladding of the Queensway Tunnel, renewal of fire mains and the renewal of high mast lighting. There are also a significant number of other smaller projects on-going on an annual basis. £12,630,000 was invested in 2012/13.

6 Claim: Other similar tunnels are toll free

Fact: Only those funded through Government as part of the national road network. There are still comparable charges for tunnels such as the Dartford and Tyne.

7 Claim: The tunnels were meant to be toll free within 10 years of construction

Fact: The local legislation, which authorised the construction of the Queensway Tunnel, may well have originally provided for them to become toll free within a defined period. However, this legislation has long since been amended through various

Mersey Tunnels Acts over many decades so that tolls remain for both tunnels to cover operating, maintenance and renewal costs and to save this burden falling on the local taxpayer. Once the debt is repaid, people in Merseyside will be consulted to determine the level of toll.

8 Claim: The surplus can be used by Mersey travel on anything

Fact: The two Tunnels are part of a wider network and cannot be considered in isolation. The surplus provides valuable revenue, which is ring-fenced for reinvestment into public transport across Merseyside as identified in the Local Transport Plan. This is what an integrated transport system is about. This is particularly important given recent funding cuts. The surplus also provides a 'buffer' should there be unexpected urgent work required on the tunnels that has not been accounted for in pre-determined capital programmes.

Given the above claim and counter claim, a similar case has existed on the Dartford Tunnel for some time and this finally reached what some will call a conclusion on 30th November 2014 when the toll became a charge (Dart Charge). You must pay for a crossing in advance or by midnight the day after crossing. You can do this using a pre-pay account, which saves you up to a third on every crossing, or by making a one-off payment. This may be in a number of ways or even via a pre-payment account, much like the fast tag on the Mersey Tunnels. Local residents need a pre-pay account to access the local residents' discount scheme.

Charges apply between 6am and 10pm and you must pay by midnight on the day after you cross. A free pass will be given if you don't pay vehicle tax because you're disabled, but you You'll still have to pay if you have a Blue Badge but aren't exempt from paying vehicle tax or you travel in a vehicle that isn't exempt.

A penalty charge of £70 will apply if you do not comply and must be paid within 28 days. It's reduced to £35 if you pay within 14 days and increased to £105 if you don't pay. In addition to the fine, you will also have to pay the crossing charge in full. Some think that this will be the way forward and the Mersey Tunnels will adopt the same or similar scheme in the future, but only time will tell and if the scheme is a success on this one.

When the tunnels are closed at night, they are normally undergoing a maintenance period to ensure that the contractors and tunnel staff can work safely within the tunnel. This would also affect the 1,400 vehicles being redirected to the other tunnel (Kingsway (Wallasey) or Queensway (Birkenhead)) compared to the 4,000 vehicles an hour during the day. Closures have to be carefully planned and suitable notice given to tunnel users. Occasionally, there may not be maintenance reasons for the closure, but filming or other similar event taking place. As you would think, evening and weekend work is more expensive, but as mentioned above but to keep the inconvenience to a minimum, this is the best time to complete the works.

There is a regular three-week cycle that involves tunnel maintenance as noted below.

Week 1
Queensway (Birkenhead) Tunnel closed for the four nights – Monday to Thursday with a 28 day notice period

Week 2
Kingsway (Wallasey) north tunnel is closed for four nights – with a contraflow through the south tunnel
Week 3
Kingsway (Wallasey) south tunnel is closed for four nights – with a contraflow through the north tunnel.

The crossing charge depends on your vehicle and account type.

Vehicle Class	Type of Vehicle	Single Journey
B	Cars (including trailers), motorhomes, passenger vans and buses with less than 9 seats	£2.50
C	2 axle heavy goods vehicles (including vans)	£3.00
D	Multi-axle goods vehicles	£6.00

Standard of Commercial Account	Local Resident Discount
£1.67	£10 a year for 50 crossings and 20p per extra crossing, or £20 a year for unlimited crossings
£2.63	£10 a year for 50 crossings and 20p per extra crossing, or £20 a year for unlimited crossings
£5.19	

You must pay for droppable axles whether they're up or down. Trailer axles aren't included but semi-trailer axles are.

You don't have to pay if you cross with a moped, motorcycle, motor tricycle or quad bike.

Strikes can have a detrimental effect on the tunnels finances and one such strike was in 1989 when NALGO called a strike, which was to cost the tunnels £200–300,000 a week in lost revenue.

At the time, it was said that Mersey travel may claw back the losses in the Poll Tax if the strike goes on any longer than it did. This was in effect a deferment of payment from the tunnel users to the overall ratepayers of the region.

During 2013, the local authorities (Sefton, Wirral, St Helens, Knowsley, Liverpool and Halton Councils) decided to combine and set up a 'Super Council' to bring greater investment to Merseyside as a whole. This would also have an effect on the tunnels. This combination of councils would be formed purely for the purposes of gaining access to major development and other such grants from central London. It would also include the tunnels coming into the group and the assets transferred into the group for the new transport authority. Halton would have sole liability for the new gateway crossing (Second Runcorn Bridge), whilst the tolls for the tunnels would be ring fenced. This super council was agreed in principle by Central Government in November 2015.

Local feelings have always run high in respect of the tolls and their ever-increasing costs and the thoughts that the tunnels should be free / part of the national road network. As you will see from the various press extracts below, this has been a long running discussion and it will more than likely continue for years to come.

Local Star newspaper an article in 1989 stated,

Merseyside MPs were pressed to lobby to end the financial burden of the tunnels. As considering the 7th rise in 15 years and with debt at £115 million, (£269,249,500.00) John Ingham, Chair of the Merseyside Passenger Authority said, "Continual pressure on DoT (Department of Transport) to cut £80 million from the debt has so far failed.

"So far as our efforts have been, the news that both the Humber and the Erskine bridges are to receive substantial financial assistance from the government. What is galling is that the case for the Mersey Tunnels is just as strong and with much wider social and economic implications.

"For example, three of Merseysides five boroughs, Sefton, St Helens, Knowsley, ae paying highly disproportionate amounts on their rates on their ratepayers make on the tunnels."

MPs are told that Merseyside ratepayers will fork out £53 million (£124,088,900.00) in grant penalties for the next ten years with costs continuing to fall on the rates well into the next century.

An article in the Daily Post newspaper on 24 February 1989, which referred to feelings running high at a meeting of the Merseytravel authority over the government's failure to adopt the tunnels as part of the national network. There were labour calls to effectively give them away along with the debt which stood at £120 million. The government have declined to say the tunnels exist as local needs despite the Merseyside Passenger Transport Executive pressing the subject. This caused Knowsley Councillors, John King, to suggest the tunnels were closed down and the resulting congestion, which would result in Cheshire County Council and the government resolving the issue.

Wallasey Labour member, Gordon Patterson, said,

"The tunnels are part of the national motorway network and it's getting to the stage where we should say take them and the debt, we wash our hands of them. MPTE have supplied the government various levels of information to the government over the last 15 years, even a detailed computer model of the option available. It is costing the average ratepayer around £600 or 9% each year and the authority will need £80 million next year.

"As a comparison, the new Conway Tunnel, which was opened in February 1992, this was deemed to be part of the national network and as such free to travel through. The main reason given for this was that the charging of tolls would have made the tunnel and diversion of traffic ineffective. It was stated that there was an existing route that is no longer than the tunnel and the whole purpose of the tunnel was to divert traffic away from Conway. I for one fail to see the reasoning for this when the Mersey Tunnels would have a better argument for being

included in the national network and as such become free to travel through if the two tunnels were to be compared like for like."

Shortly after the article of 24 February 1989 appeared, there was another from the following month detailing a third tunnel. At first glance, this may seem to suggest another cross-river tunnel but it was to be a smaller tunnel around Otterspool Promenade. This was the site of the waste disposal from the Liverpool end of the original Queensway tunnel. The residents of the exclusive area of Greensdale and Cressington Parks which holds a number of large Victorian Mansions overlooking the River Mersey, held a meeting to discuss the proposed Otterspool tunnel link. Local Civil Engineer wanted to build a new dual carriageway in a tunnel under the entire length of Otterspool Promenade. The road would go through the prom wall and sit on stilts. This never took off and the tunnel was never built. It does show that even today, despite the difficulties, there is still an appetite for pushing the engineering envelope in Merseyside.

Echo Article, Dated January 14th 2013,

THE Government is to be asked to take the Mersey Tunnels into the national road network to free the region from the burden of paying tolls.
The tunnel committee of transport authority Merseytravel voted to make the request to the Government.
It comes as the Labour-run authority is considering whether to increase tolls by 10p (from £1.50) for cars from April – a move that would generate an additional £2.6m for the authority.
Previous attempts to persuade the Government to take over the tunnels have failed and the latest bid is also likely to fail.
There is still £58m of debt outstanding on the tunnels and the business case for the planned £600m second bridge across the Mersey between Runcorn and Widnes is predicated on charging similar tolls to the tunnel.
The Department for Transport said it had not yet received the request and would consider it in due course.
The latest development follows the revelation in the ECHO that half the annual £37m collected in Mersey Tunnel tolls is diverted away from their staffing and maintenance.
*Explore our news map for latest updates from where you live
Figures obtained from Merseytravel show more than a quarter of tolls – around £10m annually – is still being used to pay off historic debts on the tunnels, which stood at more than £58m.
And another 15% of tolls, which equated to £5.6m last year, is used to subsidise other public transport services across Merseyside.
At the current rate, the debt on the tunnels is due to be virtually paid-off by 2026.
Former Wirral Council leader Cllr Steve Foulkes, who sits on Mersey travel's tunnel committee, put forward the idea of asking the Government to take over the tunnels.
He said:
"My view has always been 'if you don't ask, you don't get.
"We want to know the Government's position so we can make a decision for the long term."

He admitted the Government actually appeared to be moving towards a position of introducing more tolled roads, rather than removing fees, but that it was worth asking the question.

On February 6th 2014, the local residents opened their local press to find a story about the new toll increase for 2014. The story informed them that again tolls were to rise by 10p from £1.60 to £1.70 from midnight on Sunday 31st March. Followed was the shock to find Mersey travel had made a £10million 'profit' from Mersey Tunnel tolls. It took 14 of the 18 members of the councillors on the Passenger Transport Authority to vote for the increase, whilst the remaining four voted against. This was the second year in a row that the tolls had been increased against local pressure from businesses and the public alike.

Councillor Steve Foulkes (Wirral) said,

"This is a highly sensitive and difficult issue for Wirral members to deal with and I don't believe this is the right time to increase tunnel tolls based on the fragile economy that we're part of, and I don't feel I could support it.

Another Wirral Councillor, Les Rowlands, nominated a motion to freeze tunnel tolls for another year but this was heavily defeated.

Secretary of the Anti-Toll Group the Mersey Tunnel Users Association, Mr John McGoldrick, said,

"We think there should be a change in the way the tunnels are run, so that they're not run for the benefit of Mersey travel, and that the tolls – if we've got to have them – are no higher than is needed for the purpose of the tunnels. That would mean instead of having the toll increase, they should be knocking about 40–50p off the tolls."

Mersey travel defended the rise and argued that it was 10p less than the 'authorised' rise under the tolling mechanism, in which tolls should rise in line with inflation and stated that under this calculation, tolls would have gone up by 20p to £1.80."

Chairman of Mersey travel, Cllr Liam Robinson, said:

"Taking a decision on tunnel tolls is never an easy one. We appreciate that any increase in the cost of a service is less money in people's pockets and that's why we don't take it lightly. The tunnel tolls should rise in line with inflation each year, but we've once again kept them below this as well as offering further discounts via our Fast Tag scheme.

"The safe and effective operation of the tunnels is vital to the economy of the Liverpool city region and beyond, but maintaining the tunnels is a significant undertaking, not least because of their age. The work involves a lot more than just keeping eight kilometres of roadway usable. Millions of pounds of investment is needed to keep the tunnels on an even keel and millions more to make any improvements. This is in addition to the money needed for their day-to-day operation and to pay off the residual debt from their construction and the money borrowed for past improvements before we had the ability to raise funds in year.

"The tunnels do make a surplus. This acts as a buffer should any urgent work be required to keep the tunnels running that's not accounted for in any predetermined capital programmes."

Liverpool Echo March 27th 2014,
Mersey travel under fire over £10million 'profit' from Mersey Tunnel tolls

Accounts reveal surplus raised after all costs are taken into consideration will rise 26% next year from £7.7m to £9.7m

Mersey travel has come under fire over £10million 'profit' from Mersey Tunnel tolls.

According to tunnel accounts the surplus raised after all costs are taken into consideration will rise 26% next year from £7.7m to £9.7m.

For the first time income from the tolls will smash through the £40m barrier, rising from £38,995,000 to £41,390,000.

The revelation comes days before the controversial decision by Mersey travel to increase tolls for car drivers from £1.60 to £1.70 comes into effect in April.

Wirral councillor Les Rowlands said,

"The tunnels are a 'cash cow' for Mersey travel," and added: *"In these times of austerity, Mersey travel could have made a gesture to people who have to use the tunnels.*

"But instead they are increasing their profits on the tunnels and building up the reserves.

"Mersey travel could have raised the tolls 20 pence to £1.80 for cars – and more for larger vehicles – under the Mersey Tunnels Act 2004, but chose to limit the rise to 10 pence."

But Cllr Rowlands also stated:

"That is an extra ten pence for drivers, many of whom have to use the tunnels to get to work every day, and it all adds up.

"Just ten pence might make it sound good, but the ECHO has uncovered that they are making an increase in profits during a time of austerity.

"This is a tax on Merseyside and is affecting our economy – on both sides of the river."

The same accounts show income from tolls has risen from £37.3m in 2012-13 to £38.9m in 2013-14 and is expected to reach £41.3m over the next year.

Mersey travel said,

"The tunnel toll surplus for 2014-15 is a projected amount calculated on expected income from the new toll levels that come into force on 6 April."

Cllr Liam Robinson, chair of Mersey travel, said:

"We appreciate that any increase in the cost of a service is less money in people's pockets and we don't take it lightly.

"By law, the tunnel tolls should rise in line with inflation each year, but in keeping the tolls below the authorised levels and taking into account that discount, our Fast Tag scheme, and concessionary travel benefits, we already offer a significant discount to users in the region of £6m a year.

"The tunnels surplus acts a buffer should any urgent work be required to keep the tunnels running that isn't accounted for in any predetermined capital programmes.

"It also gets ring fenced for a funding pot that pays into a number of projects such as park and ride schemes and new bus stations – what an integrated transport system is about.

"The districts are under significant budget pressures in delivering services and so are we.

"We are getting less money from the districts so the surplus becomes ever more important in maintaining and improving the transport network for the benefit of everyone who lives and works on Merseyside."

Comments from the public (as has become the norm with the annual increase in tolls) on the latest increase, both for and against a toll rise, in the local press included:

1. Money grabbers holding the public to ransom! The toll booths are never fully open at peak times and causes mass disruption! The people must make a stand this is an outrage that cannot continue, holding people to ransom to get to work
2. Monopoly.
3. The River gives people in power the right to do these kinds of things?
4. It is essentially a border tax for getting from North Wales to Liverpool via the Wirral and it is a City/Town tax for people going to/from the Wirral.
5. We keep hearing about this surplus, but why don't Mersey travel tell us what this surplus is spent on? They justify the increase by saying we could've put it up more blah blah, but they have never told us what the surplus cash is spent on... (I don't think the shiny building at Mann Island will be mentioned). Why should they care, they are obviously not accountable to anybody and don't have to justify anything. The public need the tunnels, unless we completely boycott them, which we can't, then Mersey travel can do whatever they like, and we will be here again same time next year. It's a disgrace, if our MP's are worth anything they should grow a pair and start doing something about this.
6. How does inflation work out at a 20p increase? Inflation is at 2% and 2% of £1.60 is 3.2pence not 20p or even the 10p increase they just awarded. No wonder they can't run things right if they can't even divide by 100 and then times by 2.
7. Mersey travel have legal claims to pay out following CPO's put in place on properties required for the failed tram scheme. The rise in tunnel profits will be heading into the compensation pot for businesses / property owners impacted by the tram scheme going down.
8. £77m spent, £55m of debt accrued by Merseytravel and now more costs. Cough up tunnel users. Not for 'integrated public transport' (what new bus interchanges around the region?), but for the mistakes of Mersey travel past, and continued mistakes
9. So the tunnel is subsidising buses? I can't use buses or trains due to working in the emergency service and needing transport 24/7. Who will be subsiding my car?
10. The tunnel tolls subsidise loss-making operations on the railway and the poor strategic investment decisions of the previous CX on developing 'visitor experiences' The River made this region, but it is also a major economic

barrier as well as being a physical one. Removing the tolls, and the Mersey travel monopoly of all river crossings, would go a long way to repairing this and improving the economic viability of the region.

11. Since when has inflation (which is 2%) of £1.60 equal 20p I never know. 2% of £1.60 is 3.2p. so they should increase at a maximum by 5p as they accept 5p pieces. it generates that surplus every year so they should put away money and save it for big tunnel repairs and then get interest paid on the pot.
12. How would you feel if they reduced the price of the tunnels and didn't have this buffer then a hole appeared in one tunnel and required major money to repair it and they had to close it for good? You'd still all be moaning! We've turned into a nation of moaners. Just get on with it!

In February of 2015, it seems the tunnels had become hot property again with the general election and the constant ill feeling on tolls.

A question was asked by the local press as to: "Why are we still paying tolls?"

Within that article, it gave a breakdown to the 2013/14 financial year and the £40.2 million pounds the tunnels gained via the tunnel tolls.

Income	£40.2 Million
Capital Expenditure (Maintenance and Repairs)	£13.5 Million
Debt / Charges	£9.1 Million
Employees (Inc Police)	£4.3 Million
Supply and Services	£3.1 Million
Surplus (Merseyside Travel Projects)	£10.2 Million

As of March 2014, the debt stood at £46.8 million which included payments to (PWLB) Public Works Load Board, to cover historic costs, and modernisation and this was due to be paid off by 2048. Also included in this was an internal loan from Merseyside County Council, which was to be repaid in 2026.

In July 2015, the local long standing Member of Parliament, Frank Field, joined in the debate about the tunnels and their tolls. He called on Mersey travel to use the surplus funds to write off the debt and still allow for a substantial revenue. He stated that toll monies should be used to finance other transport projects while the debt of £46.8 million is reduced. In a letter to Councillor Liam Robinson, Chairperson of the Transport Authority in 2013/14 tolls generated £40.2 million, of which £10.2 million was used on other transport projects.

Chancellor of the exchequer, Mr George Osbourne on a recent visit to Wirral to support Esther McVey in the forthcoming elections vowed that tolls would be 'definitely' be cut if the Conservatives returned to power.

He stated,

"Surplus being generated by the tunnels was not being spent as well as it should."

A Mersey travel spokesperson stated,

"A task group is being set up to review tolling implications for Mersey travel following agreement for such at the combined authority meeting on 13/02/2015. The focus will be to consider options to reduce costs for tolls and its impact on infrastructure and transportation."

Tunnels are said to be valued at £1 billion and capital expenditure levels and maintenance paid is a modest value according to Merseytravel. It was also later reported that of the 25 million journeys made through the tunnels every year Merseytravel say that one-third (over 8 million) originate from the Wirral. As the current toll fare is £1.70 the projected annual cost of axing the tolls would be around £14 million. If the Wirral drivers used the fast tag scheme, this would fall to £11 million. In reaction to the Chancellor George Osbourne comments during the recent General Election, in that he would cancel to tunnel tolls, Merseytravel stated that a review of the tolls was underway.

Liverpool Mayor, Joe Anderson had called for a sensible approach to look at freezing the existing tolls reducing them in future years by setting up a task group. The combined authority (who now have the responsibility of tolls) task group would consider options open to them to reduce the costs. This was agreed by Wirral's Chief Executive, Phil Davies and described by Sefton's Council Leader, Peter Dowd, as eminently sensible. Mayor Anderson also called for the 2004 Mersey Tunnels Act to be reviewed within the negotiations for devolution of powers to the city region authority.

Tunnel Fraud

The morning of the 19th December 1985 was not a normal morning at the Mersey Tunnels. Both tunnels were suddenly closed by the Merseyside Police Serious Crime Squad investigating allegations of fraud and theft. Following the operation, barriers were lifted and drivers were asked to put the toll fee in honesty boxes which were positioned at the toll booths. Traffic was allowed through the tunnels approximately 40 minutes after the initial police operation was launched.

Mersey Tunnels apologised to drivers who used the tunnels and asked that they work with the tunnels in their usual helpful and co-operative manner whilst the tunnels worked through this difficult time.

At 2 p.m., Merseyside Police issued a statement stating,

A number of people are now helping with enquiries following allegations of theft and fraud.

The tunnels were closed for the evenings following the police operation and 27 people were later charged with theft and fraud before being released on bail until January 10th 1986 when they will appear before Magistrates. Once consequence of this was the shortage of staff for the tunnels forced Merseyside County Council to close the Birkenhead (Queensway) Tunnel from 10 p.m. until 6 a.m. and where needed, honesty boxes would operate at the tunnels.

At the hearing on January 10th, the hearing lasted for approximately one and half hours where the defendants were remanded on bail until a further hearing on March 7th. The men were all charged with stealing money and one had an addition two charges of handling stolen goods (24 Bottles of whiskey and a Video recorder (VCR). The defendants were charged with theft totalling £4,416.00, which ranged from £6 to £2,000.89. The local rumours of the men having purchased boats and other luxury items, which would have been beyond their earning potential and savings proved to be unfounded.

I am not going to name the defendants here or lay any allegations as that is for the judicial system but to give you a flavour of the amounts, I have listed them as follows. You will see the total amount for the defendants comes to £4063.94:

£98.80, £70.20, £141.00, £20.00, £44.40, £30.00, £25.00, £40.00, £206.95, £6.00, £300.00, £10.00, £36.40, £128.70, £2000.89, £100.00, £166.30, £103.80, £405.85, £194.60, £42.10, £56.70, £75.00, £10.00, £27.85, £82.40, £15.00.

At the hearing of March 7th 1986, the men were remanded in custody until June 2nd when a committal was expected to take place. At this hearing, a further man was brought before the courts accused of stealing £23.70 and the charges against two of the defendants were amended. One man accused of stealing £94.40 is now accused of stealing £76.20 and another man was charged with stealing £44.40 instead of £70.20. A third man was accused of stealing £10.00 now faced an additional charge involving £21.50.

During the hearing of June 2nd, the men were to be committed to trail at Liverpool Crown Court with a date fixed as July 16th. This date was inconvenient to one man due to his holiday arrangements so he was to be committed on June 26th. It was further announced that the offences date back to January 1985 with one man facing two theft charges involving £2,000 and two charges of handling stolen goods. A theft charge involving £1,600 was discharged as was a charge for handling stolen goods, namely 24 bottles of whiskey.

It was rumoured locally that some of the men had purchased luxury items to which could not be attributed to their salary, and others had unopened pay packets in draws and sideboards at their homes.

Expenditure 1970–1972

It is imperative to have a forward plan and cycle of repairs, maintenance and upgrades when looking at any large project. The Mersey Tunnels are no different and as the Queensway Tunnel has been in operation for 80 years most of the initial equipment has either been replaced or is coming to the end of its life. This can also be said for the items that have already replaced the original items. Equipment, especially in today technological work will become obsolete quickly and more efficient methods introduced. It is with this in mind that I have noted below some of the areas Mersey Tunnels have looked at in recent years.

Notably, the windows within the George Dock Building and the pumps in mid-river.

	1970-71 Actual One four lane tunnel in Operation	1971-72 Estimated One four lane tunnel and one two lane tunnel in operation
	£	£
Electrical & Mechanical Maintenance	131,000 (£1,882,299.70)	189,000 (£2,333,034.90)
Building Roadway & Tunnel Maintenance	93,000 (£1,336,289.10)	121,000 (£1,493,636.10)
Cleaning	44,000 (£632,222.80)	59,000 (£728,301.90)
Electricity	63,000 (905,228.10£)	139,000 (£1,715,829.90)
Breakdown Services	14,000 (£201,161.80)	44,000 (£543,140.40)
Administration	29,000 (£416,692.30)	33,000 (£407,355.30)
Maintenance of Property & Insurance	35,000 (£502,904.50)	40,000 (£493,764.00)
Toll Collection	153,000 (£2,198,411.10)	212,000 (2,616,949.20£)
Police Service	141,000 (£2,025,986.70)	233,000 (£2,876,175.30)
Subsidy of Birkenhead Ferries	159,000 (£2,284,623.30)	180,000 (£2,221,938.00)
Subsidy of Night Bus	4,000 (£57,474.80)	4,000 (£49,376.40)
Interest Charges and Repayment of Debts	1,071,000 (£15,388,877.70)	2,121,000 (£26,181,836.10)
TOTAL	1,837,000 (£26,395,301.90)	3,275,000 (£40,426,927.50)

*This figure includes charges for the second tunnel. The figures for interest charges and repayment of debt have been included to show the high proportion of Income taken under these headings.

(2014 equivalent amounts are shown in brackets)

Breakdown service.
The speedy removal of disabled vehicles is essential for efficient tunnel operation. The tunnel police vehicles will move light vehicles that can be towed. This equates to

approximately 80% of all breakdowns. Heavy breakdown vehicles equipment with short wave radio are maintained, one at Queensway, the four-lane tunnel and one at each end of Kingsway Tunnel during normal working days during the initial years. In today's climate, the recovery of vehicles is contracted out to local recovery companies and the tunnel police but the importance of clearing the tunnels as quickly as possible still remains. Broken down vehicles will be able to drive into the lay-by in tunnel 2B and avoid blocking one of the two lanes. It will be a shorter haul for temporary parking to allow the tunnel to be reopened or both lanes used as quickly as possible.

To ensure that all tolls chargeable are received by the Authority an automatic toll registration system is installed at each tunnel. When a driver has paid his toll in cash, the toll collector selects the appropriate button which on being pressed:

- Displays to the driver the value of the tolls paid;
- Displays the toll classification to supervisory staff, auditors and the general public;
- Signals the driver to proceed;
- Automatically records the proper amount against the collector and the tollbooth.

This system has now been updated…

Motorcycles were allowed free access to the tunnels via specified booths positioned centrally and at each end of the toll booths in Queensway (Birkenhead) but just each end in Kingsway (Wallasey).

Payments are classified and registered centrally by an electronic system with magnetic core storage. Supervisors have facilities to demand printouts relating to any collector or toll booth during a shift, in addition to the automatic print-out at the end of each shift. The six central booths have toll equipment at each end so that they can pass traffic in either direction to facilitate tidal flow, with three single booths serving each direction (13 lanes, in all). The end lanes are 20 feet wide to accommodate wide loads. The tollbooths are ventilated by the introduction of warm or cool air which escapes through the serving latch, reducing the change of vehicle exhaust gases filling the booths.

In 1966, it became evident that the Liverpool workshop, stores and garages were inadequate for serving the Queensway Tunnel. A major development notice had been served to say the site was to become part of a major development. This effectively forced the hand of the tunnel operator to find a new home for the equipment and workshops. The new tunnel enables the Mersey Tunnel Joint Committee to incorporate a new facility within the proposed tunnels and associated approaches.

The new depot houses store and workshops with garage facilities for servicing the fleet of fifty police and maintenance vehicles. Office accommodation for staff and would also include locker rooms, mess rooms and domestic facilities. The boiler house originally had oil-fired boilers providing hot water for central heating and domestic use, with a supply point for the new tunnel washing machine.

In the new Kingsway Tunnel the electrical and mechanical, technical and policing are controlled from one room, rather than the separate rooms in the Queensway tunnel. To reduce manpower the electrical and mechanical equipment are operated automatically. Carbon monoxide and visibility levels are monitored continuously at five strategic points along the tunnel roadway. The ventilation fans within the tunnel are planned to operate automatically on a time-schedule based anticipated traffic levels.

If faults develop, there is a manual override to deal with unexpected concentrations of traffic.

When the levels rise beyond certain predetermined points an automatic alarm in sounded in the control room, warning the police inspector on duty. It is then up to the inspector to make any necessary arrangements to bring the levels back down to a more acceptable level. If he feels it would be necessary, a full evacuation or part shut down of the tunnel would be sanctioned. To minimise risk of accidents and to reduce the possibility of an injured worker remaining undetected, maintenance staff work in pairs in the ventilation stations and under the tunnel roadways. These areas remain permanently lit and internal telephones are located at regular intervals. The technical section is responsible for the maintenance of both the Kingsway and Queensway tunnels, the approach roads, buildings, electrical, electronic, mechanical and lighting installations, as well as cleansing, grass cutting and even some gardening.

A vehicle has been equipped with specially designed equipment to wash the tunnel arch, walls and dados, as the specially made washing machine for the first tunnel could not at reasonable cost have been adapted to clean and new tunnel.

Various operational services have been installed to ensure safe and intensive use of the tunnel and co-ordination with the Queensway Tunnel. A control room sited over the tollbooths on the Wallasey approach provides direct observation of traffic at the tunnel entrance and is equipped with eleven TV screens and a selective desk monitor. Traffic within the tunnel is safeguard by monitoring of atmospheric CO_2 and visibility at key points, and by provision of emergency points every 160 feet (48.7 metres) along the tunnel accommodating alarms and telephones to control, hose reels, fire hydrants, and chemical extinguishers. There are also radio links with police and tunnel service vehicles and with the Queensway Tunnel.

Ventilation, lighting, pumping and main and standby electricity supplies can be operated by remote control if it is necessary to override automatic controls. Traffic controls are coordinated with those of the Queensway Tunnel and with Liverpool City and the Mid-Wirral Motorway.

Mersey Tunnels Expenditure 1970-73

	Operating One 4 Lane Tunnel		Operating One 4 Lane & One 2 Lane Tunnel			
	Actual Cost £1,000 1970-71		Probable Cost £1,000 1971-72		Estimate Cost £1,000 1972-73	
Electrical and Mechanical Maintenance	0.0	130	0.0	159	0.0	192
Building tunnel & roadway maintenance		81		124		134
Cleaning		45		58		64
Lighting & Power		63		107		127
Breakdown Services (less receipts)		13		19		19
Administration		33		33		33
Insurance, maintenance & property miscellaneous expenses		36		54		50
Toll collection		161		209		214
Police		140		216		228
Subsidy Birkenhead ferries		170		220		205
Subsidy night bus		2		2		2
Debt charges						
Birkenhead tunnel	76		76		76	
Wallasey tunnel	573		1620		2302	
Birkenhead approach roads	414	1063*	475	2171*	505	2883*

Total expenditure		1936		3372		4152
Total income		1937		2850		3150

*These figures include interest charges and repayment of debt for the second tunnel and have been included to show the high proportion of income taken by department charges.

As with any structure particularly tunnels and bridges, there is always an on-going assessment. As the Queensway Tunnel approached its 80th birthday and there were numerous legislative and other changes, not to mention the wear and tear on the equipment and materials, it will more than likely all have to be replaced.

During the 2002 Cross passage works within the Kingsway (Wallasey) Tunnel, Mersey Tunnels took the opportunity to save on costs and to undertake some additional works for the Lay-By. This had been suffering for a number of years from leaks through the cast iron tunnel lining and this had badly corroded secondary lining, so was in need of some serious care and attention. The cross passage works would see the tunnels closed and therefore an ideal opportunity to carry out the required remedial works on the lay-by.

When the tunnel was constructed, steel sheet secondary panels, with an asbestos backing, were installed and had a PVC plastisol paint coating to the traffic face. Once the dado panel at the base had been removed, there were obvious signs of deterioration of the panels, which had gone unnoticed for many years. The work would see the secondary panels removed, and be replaced with non-asbestos backed panels. This would also be the case for the dado panels, which would have a safe cement based backing and an overall refurbishment and re-installation.

As work progressed, it became obvious that it was worse than initially expected, and some corrosion of the supporting steelwork had also taken place. It was there for decided that a complete refurbishment of the lay-by would be the best option, which at first hand may seem a little extravagant, but in the long term would be a saving in maintenance and safety of the people using the tunnel. The new secondary lining would be of a ceramic-coated steel type with the additional work of cleaning and re-painting the original tunnel segment lining to increase its life expectancy. Most of the works were completed by August 2002.

2011/12 Planned Programme of Maintenance and Medium to Long Term Investment Plan for the next 10 years.

The overall outturn forecast for the 2011/12 capital programme expenditure was estimated at £9.2m. The main scheme delivered this year has been the renewal of the Queensway Tunnel Cladding which was completed in January 2012.

In addition, successful completion of the Kingsway tunnel Fire mains was completed. In total, there were 53 projects successfully completed and a further 6 schemes will continue into next financial year.

The 2012/13 Capital programme has been built up in a similar pattern to previous years.

Tunnels PA system

Queensway Tunnel Cladding – Scheme detailed above and completed in Jan 2012 at a cost of £7.05 million

Bidston Moss Viaduct Works Contribution

Kingsway Fire Main Update – Renewal of two main fire mains within Kingsway Tunnels which are over 40 years old and increasingly failing causing local flooding and loss of fire mains pressure. The scheme was completed at a cost of £3.5 million, which is below; the estimate as own internal labour was used to support the contractor.

Mersey Tunnels CCTV – Scheme was to renew the Tunnels CCTV and development of a new central storage for CCTV images at a cost of £1.0 million

Kingsway Concrete Slab Repairs

Fire System Suppression System Works

Sherlock Lane Footbridge

Instal Roof Solar Panels – Scheme to generate 50Kw of energy to cost £150,000 with a potential saving of £50,000 a year in energy costs.

Victoria and Promenade Ventilation Stations Refurbishment – External refurbishment and other works which included concrete repairs, flat roof renewal, masonry repairs and rainwater goods.

Woodside Scaffolding to test Lifting Beans and New High Level Safe Access Walkways

Invert/Station Phones – The current system is almost 25 years old and due to the harsh environment of the tunnels the system has become maintenance intensive.

Access Controls – Cost of new access system £200,000 over two years

Air monitoring Systems – New system to measure Nitrogen Oxide as well as existing systems measurements for Carbon Dioxide, and air particle levels. Scheme to cost £240,000 which is a saving from the original estimate of £500,000 following a competitive tender.

Renew Kingsway Approach Road Lighting (Wallasey) – Cost of works £25,000

New Windows at George Dock Building – Existing 1934 windows within the listed building to be renewed due to end of life and a majority are unable to be opened. Scheme to cost £400,000 as they have to be refurbished rather than replaced which would have cost £250,000

Kingsway South Tube Renewal of Walkway Decking

Inspect and Repair all fixed ladders at all Stations and Inverts

Lime Kiln Lane Switchgear – Renewal of 415 Volt switchgear as existing equipment is 40 years old

Wallasey Workshop Works

Overhaul Promenade Ventilation Station Fans

Radar Control Speed Signs

Renewal of Roof to Old Haymarket Police Point – Cost of works £8,000

Queensway HGV Warning System – Working closely with Liverpool City Council to provide a new warning system to prevent HGV's entering the Queensway Tunnel.

Bus Beacon System – An estimated £30,000 is to be spent on this Real-Time Information (RTI) system to provide a number of beacon stations within the tunnels to monitor buses passing through the tunnel.

Mandatory H&S Schemes

SCADA – (Supervisory Control and Data Acquisition) A separate system for each tunnel to monitor fans, pumps, and lighting.

Combined Control Room – To be deferred to 1212/13 due to new HQ at Mann Island

Mersey Tunnel Fire System Upgrade
Ascom Phone System renewal
Integration of New Phone system, Radio Channels and Hardware
UPS Upgrades and Battery Replacement (Uninterrupted Power Supply)
Speed Warning Signs on Wallasey Approach – Traffic calming at Wallasey Approach – It has been observed that some drivers are approaching the toll barriers to quickly and a system of rumble strips have been installed along with new signage

Priority H&S Schemes

- Kingsway North Tube Walkway Decking Renewal
- Kingsway Liverpool Side Approach Booster lighting renewal
- Queensway Approach Booster Lighting Renewal
- Kingsway Pump Motor Control Panel Renewal – There are three main pumps located in mid-river and are the original ones installed in 1934. As these pumps are still in production, they have been replaced proposed to renew them on a like for like basis
- Replacement of Bollards at Toll Lane Noses in both Queensway and Kingsway

Prioritised New Starts for 2012/13

- Queensway Tunnel Road Deck Resurfacing
- Customer Information Signs
- Kingsway Tunnel Renew all Wall Panelling and Finishes (Feasibility Study)
- Replacement of High Mast Lighting Liverpool, Wallasey and Birkenhead
- Inspections and Assessments
- Various Tunnel Structures
- Borough road flyover – Structural Survey
- TERN – Tunnel Assessment if Kingsway is designated TERN (Trans European Road network) tunnel
- Investigation of Kingsway and Queensway Ventilation systems
- Kingsway and Queensway road Deck Inspection and assessments
- Consultation on new Toll Collection methods
- CCTV Incident Detection
- Queensway Tunnel Drainage System (Feasibility Study)
- Ventilation Station Fan Overhauls
- Hinson Street – Replacement of Asbestos roof and Strengthen Floors
- Hinson Street – Access Hoist and Goods Lift
- Building Refurbishments
- George Dock Entrance Doors
- Replacement of Grid Covers and drains in Kingsway Plaza
- Woodside ventilation Station Windows
- Vehicle Procurement
- New road tunnel cleaning machines. Existing machine is 10 years old and suffers reliability problems. A new machine has been purchased which is based on a Mercedes Unimog and has been fitted out by Mulag a specialist tunnel cleaning company

- New transit vans – At least one of the existing fleet of Transit vans has become beyond economic repair following a safety check. Two new vans were delivered in March 2012

Electrical Renewals

Coin Handling Machine Software Upgrade – A fully automated vacuum system moved the coins from the toll booth to the cash rooms. Recently these have been breaking down and the coins have had to be moved manually. The Queensway was the first to try for an upgrade followed by the Kingsway by utilising the valve operation sequence.

Medium and Long-Term Plans

- Safety
- Customer Needs
- Ensuring Reliable Operations of tunnels
- Increase the Tunnel Capacity
- Reduce Future Maintenance Costs
- Improve Quality of Infrastructure

Planned Expenditure Medium to Long Term
(Source Mersey Tunnels Delivery Capital Programme 2011/12 Programme 2012/13 and 10 Year Look Ahead Programme)

Scheme	2012/13	2013/14	2014/15	2015/16	2016/17	2017/18	2018/19	2019/20	2020/21	2021/22	Totals
New Scada System		£1,500,000	£1,500,000								£3,000,000
New Combined Control System		£500,000									£500,000
Renew Kingsway Tunnel Wall Cladding		£4,000,000	£4,000,000	£3,850,000							£11,850,000
Kingsway Tunnel Replace High Mast Lighting		£1,500,000	£1,000,000								£2,500,000
New Toll Collection System Kingsway and Queensway				500,000	£4,000,000	£4,000,000	£1,000,000	£3,000,000			£8,500,000
Remodel Kingsway Toll Plaza (Demolition of Elevated control room)				£250,000	£1,000,000	£1,000,000					£2,250,000

Item							Total
Remodel Queensway Toll Plaza					£500,000		£4,500,000
Kingsway Tunnel Road Surfacing - Low Rumble Surface	£1,000,000						£1,000,000
Rewire Queensway Tunnel		£500,000	£2,000,000	£3,000,000			£5,500,000
Rewire Kingsway Tunnel					£500,000	£4,000,000	£4,000,000	£8,500,000
Kingsway - Highway Improvements Scotland Road					£500,000	£2,000,000		£2,500,000
Replace Sump Pump Drainage pipework System Queensway Tunnel	£250,000	£2,000,000						£2,250,000
Replace Sump Pump Drainage pipework System Kingsway					£500,000	£2,000,000		£2,500,000

Item	Y1	Y2	Y3	Y4	Y5	Y6	Y7	Y8	Y9	Total
Tunnel Queensway/Birkenhead Flyover Maintenance Works		£1,500,000								£1,500,000
Resurface M53 Approach Bidston Moss Flyover				£1,500,000						£1,500,000
Renew Jet fans					£1,000,000					£1,000,000
Vent station interior refurbishment programme	£500,000	£500,000 3333	£500,000	£500,000	£500,000	£500,000	£500,000	£500,000	£500,000	£4,500,000
Vent Fan Refurbishment programme	£250,000	£250,000	£250,000	£250,000	£250,000	£250,000	£250,000	£250,000	£250,000	£2,250,000
Lift Renewals			£500,000	£250,000			£250,000			£1,000,000
Bridge Infrastructure Works	£500,000				£500,000					£1,000,000
Update of Tunnel Estate H&S Document	£100,000	£100,000	£100,000							£300,000

ation/Asset Database									
Vehicle Replacement Programme		£250,000	£250,000	£250,000	£250,000	£250,000	£250,000	£250,000	£2,250,000
Highway Surfacing Works		£500,000			£250,000	£500,000			£1,500,000
TOTALS	£10,512,000	£10,850,000	£10,100,000	£9,700,000	£9,500,000	£9,750,000	£11,500,000	£8,500,000	£1,250,000	£1,000,000	£82,662,000

To look at a few of the updates in depth, to enable the reader to ascertain the work involved in what at first seems like a basic maintenance project.
Much of this work goes unseen during annual maintenance periods or the odd waggon that is seen on the roadside. A lot of the infrastructure is houses is special buildings, such as the ventilation buildings. All of the work is organised by the Mersey Tunnel Asset Management Department who are responsible for all buildings within the organisation.

New Wall Panels at Queensway Tunnel

The current wall cladding which was installed in 1983 by Merseyside County Council no longer complies with safety regulations. As with most public buildings, and authorities, they have to comply with stringent additional regulations in respect of public safety. The original panels had been white in colour, had discoloured and become yellow over the years. This staining and discolouration was also proving more difficult to remove as it gained a foothold into the material.

In the 2010 maintenance report, the Mersey Tunnels stated that they were looking at re-Cladding the inner tunnel lining. The panels are to be replaced by Bahrain Fibre Glass (BFG) for the sum of £7,059,801 and the cost is to be spread over two years. A local company, Tarrin Engineering, which has undertaken work in the tunnel before are to install the panels along with MDE Projects of Essex. The framing for the new panels were supplied by Ancon Building Products in Sheffield.

Within the new panels, Merseytravel have commissioned a work of art to be placed at the end of the tunnel, along with new coats of arms for Wirra and Liverpool. The coats of arms are to be placed on the boundary of the boroughs within the tunnel. The work of art was designed by Alison Barker after winning a competition by the Public Arts Steering Group. The art work will take up 18 of the 1.2-metre-wide panels at either end of the tunnel. What makes this more significant is that this was the first road tunnel in Europe to house an art installation.

The piano keys of the Liverpool entrance depict the City's musical heritage and the green foliage at the Wirral side depicts its historic plants from which the name 'Wirral' came from. Both panels will also show a sky line of their respective river fronts.

The 4700 panels proved to be a task as they were 5.1 metres in height and over 1000 reduced height panels in the Liverpool Dock branch at a height of 4.0 metres. In addition, they have to work around the 82 fire boxes, various ventilation Emergency Refuges and CO_2 monitors within the tunnel.

Kingsway Tunnel Fire Main Renewal

The fire main, which as you will imagine is a major integral part of the tunnels safety system. Due to the nature of the tunnel and its atmosphere, the fire main has badly deteriorated. This deterioration caused breakdowns in the system and flooding from leaks, both of which have caused a loss of pressure to the fire system. A £3.5 million scheme was put together to renew the system and ensure the safety of the tunnels

The new fire main system is designed as a ring main supplied at each of the two tunnel entrances and ventilating station buildings and connected into the external mains network operated by United Utilities. The main comprise of approximately 150 mm diameter 1.8 km length buried High Performance Polyethylene pipe (HPPE) section and 4.5 km stainless steel pipe section running beneath the road sections in the airways (invert sections) of the north and south tunnels. The main is connected to 75 mm diameter fire hydrant points (94 units) evenly spaced under the walkways on both sides of the carriageway

The new scheme design has allowed for the relocation of the tunnels fire main to the fresh air chamber beneath the road decks for easy operation and maintenance. This will ensure a reliable system with a minimum maintenance for at least 30 years. The system will incorporate active flow monitoring, leakage detection and Control to reduce water loss.

Sherlock Lane Footbridge

A structural survey of the existing concrete footbridge structure identified that it was in poor condition and required renewal. The project required the demolition of the existing footbridge within a weekend closure of the four lanes of the approach road in co-operation with the Highways agency. Due to the weight of the two main concrete beams spanning over the approach road two cranes were required. Once demolition works had been carried out the bridge pier foundations were refurbished prior to the delivery of the new bridge. The new design was a single span steel bridge with mesh sides was craned into place over the weekend and within original budget of £170,000

Wallasey Depot Works (New Fuel Tank)

A new 20,000-litre fuel tank (bundled and double skinned for total integrity in the event of a leak) and associated replenishment controls were installed in six locations, covering each of the emergency power generators. Each tank has been fitted with high and low alarms and electronic contents gauging to allow central monitoring. The operations centre vehicle fleet fuelling forecourt was also upgraded. This included relocation of the dispenser pump and the installation of a surface drainage interceptor to prevent oil spills and contaminated water from entering the storm drains. The pump was modified to allow remote monitoring of all fleet usage and stock levels

Georges Dock Asbestos Removal

Mersey Tunnels took the decision to remove the asbestos from the six ventilation shafts within the George Dock Ventilation building. This asbestos was installed around 1965 as sprayed on fire cladding to the structural beams in the mid-river fan chamber. The cladding had become unsafe and prone to accidental damage thus accidentally releasing asbestos fibres into the tunnels and tunnel tour route.

The beams were cleaned down, coated with corrosion resistant paint and new fire resistant cladding applied. All hazardous waste was disposed in accordance to the Hazardous Waste Regulations 2005 and reusable items were recycled. The scheme was completed to time and programme in February 2011 at a cost of £89,000

Woodside Gearbox Vent Inspection

The Queensway tunnel is ventilated from six vent shafts in Liverpool and Birkenhead. During a routine maintenance inspection, the gearbox was one of the original pieces of equipment fitted in 1934 and the continued running is a testament to the equipment chosen and the maintenance routing of the Mersey Tunnel specialists.

Capital Programme Works 2014 / 2015

DETAILS	TOTAL	SPEND TO 31/03/13	REVISED ESTIMATE 2013-14	DRAFT ESTIMATE 2014-15	DRAFT ESTIMATE 2015-16	DRAFT ESTIMATE 2016-17
	£000	£000	£000	£000	£000	£000
Tunnels Ongoing Schemes						
Renewal of Air Monitoring Sensors	145	137	8			
Renew Kingsway Approach Road Lighting (See story below)	132	122	10			
Victoria & Promenade Vent Station Refurbishment	318	293	25			
Queensway Tunnel Road Deck Renewal	2,048	1,992	56			
Replacement of Highmast Lighting (L'Pool, B'Head, Wall)	1,696	396	1,300			
GDB Window Replacement	432	210	222			
Woodside Internal Building Refurbishment	578	378	200			
Woodside Vent Station Window Renewal	252	2	250			
Automatic Incident Detection	524	274	250			
Vehicle Procurement	677	281	396			
Ascom Phone System Renewal	799	599	200			
Kingsway Pump Motor Control Panel Renewals	231	106	125			
Highway Grid Repairs – Wallasey Plaza	100	40	60			

293

Toll Collection Feasibility Study	52	0	52
Kingsway (L'Pool Side) Approach Booster Lighting	50	0	50
Queensway Tunnel HGV Vehicle Warning System	43	23	20
Queensway Booster Lighting	40	0	40
Access Control	157	117	40
Invert Telephones	246	216	30
Kingsway South Tube Renewal of Walkway Decking	104	79	25
Building Refurbishment	169	154	15
Electrical Renewals	33	18	15
Kingsway North Tube Renewal	107	81	28
Integration of New Radio Channels	165	102	63
Queensway Tunnel Drainage Feasibility Study	41	26	0
Kingsway Resurfacing &Traffic Calming	62	58	4

An article in the Liverpool Echo gave the details on the new lighting works for the Wallasey Tunnel. These works were to include 40,000 metres of electrical cable along with 1,700 LED (Light Emitting Diodes) lights over a two-year period. The work will include relocating cabling from along the side of the tunnels to beneath the road, reducing the chances of power being affected by a crash and making maintenance easier. The new lighting will also be more energy efficient, cutting CO_2 emissions by around 338 tonnes a year and saving £66,000 in electricity costs.

The rewiring is part of a 10-year investment programme for the two tunnels to keep them in order and able to accommodate the 25 million vehicles which pass through them each year. The work will be funded through money set aside from the tunnel tolls.

The £8,000,000 scheme will be carried out under planned closures at evenings and weekends to allow the tunnels to function normally through the week and will be the first actual replacement of the lighting since the tunnel was opened.

Frank Rogers, Mersey travel's deputy chief executive, said:
"The tunnels are a vital part of our transport infrastructure, providing important road links for millions of vehicles each year, helping to sustain and grow the economy of the region.

"They are amongst the safest for their age in Europe and it's essential we continue to invest in them so we can keep them up and running for generations to come."

Despite the millions of pounds spent on the tunnels and their infrastructure, the tunnel still has day-to-day problems, as with any structure, particularly those in their early 80s.

Some of these problems are not of the tunnels making and stem from outside sources but they can still have a significant effect on the tunnels day-to-day life.

Wallasey Tunnel Asbestos Removal

In recent years, the area below the road deck of the Kingsway tunnel has had the asbestos removed. This involved a team of specialists raking out and replacing the asbestos joints with 'Fastfill' which was followed by 'Cemprotec E942' and reinforced with 'Cemprotec Scrim' to seal the joints and lock in any residual fibres.

The reason for this work was the tunnel was originally lined with pre-cast segments, which had the caulking material containing asbestos. At no time, would the users of the tunnel been in any danger due to its location and the highly regulated and controlled way the caulking containing asbestos elements were removed by a specialist company.

With the passage of time and small amounts of water ingress, the fibres breakdown in the joints and it was thought that this would best be removed. The works were carried out by Flexicrete Technologies Ltd of Leyland.

Queensway Tunnel Flooding December 1991

Queensway tunnel was flooded due to a burst water main in William Brown Street (opposite the tunnel entrance) and 500,000 gallons of water ended up in the mid river section, which is the lowest section of the tunnel.

With the tunnel pumps dead, Merseyside Fire Brigade was not able to pump the water out so additional pumps were hired and the water pumped out into St Georges

Dock pump room and then on into the River Mersey. The tunnels were closed for the entire weekend whilst the pumps were refurbished and a new electrical control panel was fitted due to excessive water damage.

Queensway Tunnel Flood, Monday 30 December 2002

As the tunnels get older, their maintenance becomes more important and with it comes the additional expense associated with the care and maintenance of old buildings. Following a failure of one of the tidal pumps, the Birkenhead Tunnel was closed as it began to flood with water from the River Mersey. Fire crews from Merseyside Fire and Rescue were called to pump room Number One at the Queensway entrance in Birkenhead at around 7 am when the water level reached 2-ft and threatened nearby electric wiring.

As a precaution, and to allow the emergency services along with Mersey Tunnel staff to deal with the situation safely, the tunnel was closed and traffic diverted to the Wallasey (Kingsway) Tunnel. Fortunately for the time of day, the traffic as light but coming up to rush hour, this would cause some traffic problems in the surrounding areas. Upon initial investigation, it was suggested that the pump valve was blocked and thus caused the pump to short circuit. After a short time, the fire brigade had removed the water from the pump room to allow temporary repairs to be made.

Later that day engineers from Mersey Tunnels visited the site to ascertain the damage caused and the necessary remedial action and repairs required.

23rd February 2016 saw the announcement by Liverpool City Council that Liverpool is proposing to pay £2 billion of the 33 billion cost for the new HS2 rail line to have a branch to Liverpool. This was followed by an article in the Liverpool Echo stating the funding should come from the Mersey Tunnels. You may understand that the local feelings on this were to say the least, not impressed and that the tunnels were again being used as a local tax cow.

A local think tank, ResPublica, stated that,
"The HS2 link for Liverpool is essential if Chancellor George Osborne's Northern Powerhouse plan is to succeed and Liverpool is not to be left behind and the 'superport' Liverpool 2 is to grow."

Liverpool Mayor, Joe Anderson stated that,
"Using this mechanism would allow both ourselves and the wider economy to move closer to prosperity."

A report by ResPublica entitled *Ticket to Ride: How high speed rail for Liverpool can realise the Northern Powerhouse* calls for city region to link into HS2.

The report has been assisted by Grant Thornton and it was calculated that the current capitalised value of this surplus of the tunnels is around £500m over a 35-year period.

Wallasey MP Angela Eagle said,
"The report makes a strong case for investment in the rail link and it rightly highlights the strong economic performance of the North West in recent years, and its potential for further growth in the future."

However, Wirral council's leader, Phil Davies, has said,

"That Mersey Tunnel tolls should not be used to help fund a high-speed rail link into Liverpool."

Although, 'strongly supporting' a high-speed rail link to the city region, he stated, "My view remains that the surplus on the tunnels should continue to go towards bringing down the cost for Wirral residents and this is something we should look to George Osborne to fund. The big priority for me as leader of Wirral council – and that's one of the cornerstones of the devolution deal we signed with the Government in November."

He also stated that they had made a good start on sorting out the tunnels with the decision this year to freeze cash tolls and reduce tolls for people using the prepay Fast Tags. He added: "I would be wary of signing up to another complicated financial deal that would tie us into interest payments over the next however many years because that was the deal when they were built all those years ago and we're still trying to extricate ourselves from that."

Queensway Toll Point (Author)

Chester Street Exit Toll Booth (Author)

The following four images show the extent of the constant maintenance required on the 1969 flyover scheme from the A41 to the tunnel tolls.

Most of the areas are showing examples of ongoing maintenance requirements. The images were taken in summer 2015

Sections of Stone Facing to Concrete Flyover are Coming Away from the Concrete Base

Sections of Stone Facing to Concrete Flyover are Coming Away from the Corner of the Concrete Base

Close up of the Joint within a Box Section

Section of Flyover Showing the Joint within the Box Section

Tunnel Walls on Liverpool Dock Exit Once

New Damaged Cladding has been removed for Repair 2016. (Author)
The water Seepage is normal for any tunnel and this section is just under the Atlantic Hotel on the Strand.
You can see the new cladding brackets at the bottom of the photograph

New Salt Barn at Queensway (Author)

Chapter 17
Second World War

Between 1939 and 1945, the port of Liverpool handled more than 75 million tons of cargo, mostly from the Atlantic Convoys. 56 million tons were imports vital to the war effort and feeding the public along with 18.5 million being sent back out as war supplies to the battle fronts throughout the world including North Africa and Main Land Europe. At the start of the war, many proposals were being considered for storing and using many materials for the war effort and one of those was the Mersey Tunnel.

To allow for all this work, the War Office took great interest in the Mersey Tunnels In a letter of August 24th 1940 from Sir William G Chamberlain of the Ministry of Transport Office in Manchester to Sir John Reith GCVO, GBE MP of the Ministry of Transport in London. In this letter, the Manchester Office was reporting on the survey of the Mersey Tunnels to ascertain their possible use for the war effort.

A proposal of a number of alternatives were detailed in the letter as follows:
As I intimated to you on the telephone, to close the Tunnel entirely too vehicular traffic would create a situation in Liverpool which would not only be difficult but which would be chaotic. Apart from commercial road transport with which I am more closely concerned, the effect of military traffic present and prospective would of course need to be taken into careful account but I gathered from your conversation that you had this aspect in mind. The only point I would like to mention is that the movement of troops, munitions and stores in the event of a major emergency arising would be a matter of the utmost urgency.

I have, however, carefully considered possible alternatives which would in the main provide for the partial use of the Tunnel for purposes of road transport and also partially for the purpose of National Defence. My own knowledge of Liverpool traffic, which goes back ten years since my appointment as Chairman of Traffic Commissioners, convinces me that to close the tunnel entirely to road transport for the duration of the war would cause grave traffic congestion. And in my view the facilities provided by the river ferries at present could not in any way deal with the volume of traffic which passes from one county to the other. Although, I am aware of the extent of space which you may require for national defence purposes I think you may find that one of the alternatives described below will provide for the purpose you have in mind without suspending communications.

Alternative 1

To close two of the four tunnels lanes throughout the whole length of the tunnel. This would give a distance of 2.01 miles and a width at the widest point (road level) giving a clearance of approximately 22 feet. Regards must be had however for the curves in the tunnel at certain points as will be observed from the general plan enclosed.

By this arrangement two lanes would remain open for road traffic would receive the inward flow from two lanes at each end i.e. one inward from the ranch tunnel and one inward from the main tunnel at each end. After passing the junctions proceeding inwards, slow and fast traffic would then have only one lane in each direction thus slowing up the traffic generally to the speed of the slowest vehicle. The traffic capacity would therefore be curtailed but to offset any possible congestion it would presumably be practical able to exclude from the tunnel private motor vehicles except with a claim to priority.

Alternative 2

Use the branch tunnel for your purposes. These two tunnels are of two lanes with a width at road level of 26 feet 6 inches approximately. The total length you would then have for your purposes would be 3,218 feet that is 1,564 feet at the Liverpool and 1,654 at the Birkenhead end.

This arrangement would of course have the effect of diverting traffic on the Liverpool side from the New Quay (Dock Entrance to the Queensway (Old Haymarket entrance and on the Birkenhead side from the Rendel Street entrance to the King's Square (Chester Street) entrance. I have not had an opportunity of discussing the question of the transference of this traffic on Birkenhead side with the Birkenhead Corporation representatives but from my own knowledge of the district, I do not think this would create any surmountable difficulty.

On the Liverpool side, however, the transference of traffic normally using the New Quay Entrance to the Queensway Entrance would increase considerably the volume of traffic in the centre of Liverpool. But in view of the fact that the major part of the goods traffic originating along the line of the docks generally could make its way to the Queensway Entrance by may alternative side roads, I am of the opinion that any serious traffic congestion could be avoided. You will appreciate that by this arrangement the whole width of the main tunnel would remain open for vehicular traffic.

Alternative 3

To close that portion of the main tunnel, which lies between the Queensway Entrance and the junction of the main tunnel with the New Quay branch on the Liverpool side and to close the portion of the main tunnel which lies between Kings square and Rendel Street branch on the Birkenhead side. In my view, this would create considerable congestion at the New Quay Entrance in Liverpool. This entrance is already a point dealing with a considerable volume of traffic, and to transfer to it all the Queensway traffic would create a chaotic positon.

Alterative 4

To close two lanes of the four at each end of the main tunnel from the entrances to the junctions.

This arrangement would leave four lanes available in the main tunnel between the junctions, would leave all four entrances open, but would cut down the four lane entrances at Queensway and Kings square to two way. In other words, it would give four lanes through the tunnel but bifurcating at the junctions. It may be found possible to cater for all the existing traffic by this method without having to place restrictions upon any particular type of vehicle.

Although, I have noted some of the dimensions which you gave on the telephone this morning, there are other dimensions such as height, of which I have no knowledge but it may be a useful suggestion for those concerned if it is intended that the planes are to be housed with their wings attached for consideration to be given to the question whether housing the aircraft sideways may not be practicable if normal housing is not feasible. Small trolleys of bogies fixed at right angles to the plane wheels would in that event have to be used.

You will observe from the plans that for a distance lying under the waterway there is a tunnel below the roadway. This is termed the 'future traffic tunnel' and the original intention was to use it for tramcars with a breakaway at each side of the river remote from the existing entrances.

I have discussed this with the Liverpool City Engineer the question of access to this future traffic tunnel but so far, as I could ascertain, it would not be practicable to use it except for stores or equipment in cases. The existing access to the future traffic is only by way of 3 feet by 6 inches entrance.

The City Engineers said, however, that he would be prepared to consider the practicability of cutting through the tunnel roadway to gain access to the space below. The situation of the space is shown clearly on the accompanying plans; the dimensions are 17 feet in height at the highest point and 21 feet in width.

At the foot of the letter, a note has been made stating, "Maybe alternatives laid together."

You will remember that when you spoke to me this morning you mentioned a 33-foot road on the north side and a 36-foot bridge on the south side. I have not been able to trace which particular road and bridge you have in mind but I have examined the position on both sides of the river in conjunction with the City Engineer ad a representative of the Divisional Road Engineer. The city Engineer said that even if there were a road of only 33 feet, which presented difficulty, there would doubtless be alternative ways of approach. The representative of the Divisional Road Engineer stated that if the span of the lanes is in fact 36 feet 10 inches there would be many other roads on the approaches to both sides of the tunnel (though not in its immediate vicinity) which would be narrower than 33 feet

At the present time, the Liverpool City Engineer is arranging for the conveyance from the docks to Speke Aerodrome of completed American made planes. He described them as medium bombers and those of the fighter class. Their conveyance by road necessitates special arrangements being made such as the stoppage of traffic at certain times of the night but in view of the fact Gladstone Dock and the dock road leading to the tunnel is of ample width, I convey this information to you in case it is contemplated that the aircraft in the tunnel are those arriving in Liverpool from America.

If there is any further information you require, you will no doubt let me know. In the event I shall be of available at the office here in Manchester tomorrow, Sunday from 10 a.m. to 5 p.m.

As I have borrowed the plans from the City Engineer, you will no doubt let me have them back when you have finished with them so I can return them to him.

Apart from the plans there were also nine pages of details showing the tunnel traffic and cost attached to the letter.

A letter to 'The Minister' dated 27 August 1940 marked SECRET:

This letter refers to the letter of August 24 1940 from Sir William G Chamberlain of the Ministry of Transport Office in Manchester to Sir John Reith GCVO, GBE MP of the Ministry of Transport in London.

The letter outlines the arguments set out in the letter from Sir William and also notes that there is no mention of the possible danger of fire which was always uttermost in the minds of the ministry, particularly with aircraft and the risk from either accident or sabotage, especially in such a confined space. The letter notes that it would be extremely difficult to deal with this sort of situation if the tunnel were packed with aeroplanes, no matter how much the existing fire appliance were supplemented. It would be for this reason and the contraction of traffic facilities alternative 1 would not be looked into.

Alternative 2 was considered above alternative 3 by reason of the fact that New Quay and Rendel Street entrances were less used then the main entrances. Alternative 4 would also be considered as this was seen to have a 'clean cut' between those portions of the tunnels used for traffic and storage. If aeroplanes were to be the essential storage choice of the tunnels, then alternative 2 would be the preferred option. This would see the two entrances being blocked at the point of their junction with the main tunnel and the use of fireproof doors so as to provide two means of entry and exit.

As for the mention of a space under the main road tunnel, it was considered that the effort required to access this space would be too late in the day talk of a 'breakaway' from this sub tunnel on each side of the river. The proposal did however warrant further investigation as to the use of hoists and access through the tunnel floor. If aircraft were to be stored in the tunnel, then they may have to enter and leave the tunnel at each end of the main entrances, otherwise a great deal of space may be wasted which may go against using the tunnels for aircraft storage.

It was proposed to consult with the air Ministry to find out exactly what aircraft it was proposing to store in the tunnels.

Notes from this letter and the accompanying plans were sent to

The Rt. Hon the Lord Beaverbrook, Minister for aircraft production and The Rt. Hon Anthony Eden, M.C. M.P., War Office

The war office replied with their observations on 28th August via Mr O'Neill (Mr Eden's Secretary) and Mr Eden was away in the country. The General Staff had been consulted on the proposals and it was thought that the proposals, given the tunnels geographical location and proximity to Widnes and Warrington, notwithstanding the daily use of the tunnels, that two-way traffic should be maintained as this was the main arterial route between Birkenhead and Liverpool. In a further letter, it was noted that the tunnel was used at night for heavy RAF and Admiralty traffic with some of the RAF vehicles being 18 feet wide and as such would not be able to use the tunnel if two roadways (traffic lanes) were available.

On September 24th 1940, Sir William G Chamberlain of the Ministry of Transport Office in Manchester to Sir John Reith GCVO, GBE MP of the Ministry of Transport in London informed Sir John Reith on his previous days meeting with Alderman Shennan, the Leader of the City Council and Chairman of the Tunnel Committee, Mr Baines, The Town Clerk, and Mr Hamer, the City Engineer.

Sir William noted that during the meeting it was obvious that the views of the Corporation had somewhat changed, possibly as a result of the extensive bombing the city had experienced.

Alderman Shennan made the following four points:

1. That the people of Liverpool would raise objection to the use of the tunnel for Air Ministry purposes when they have been refused its use as a deep shelter from bombing.
2. That the tunnel legs, by virtue of their gradients and ventilation and sanitary arrangements would be quite unsuitable for the use as a manufacturing workshop. (This referred to the proposals by Vickers Armstrong Ltd to use one of the legs as a manufacturing base for Wellington bombers with the part delivered to Blackpool and Chester for assembly)
3. If the legs were adapted for workshops, the Liverpool aircraft establishments, namely Rootes securities and Napers who manufacture bombers, ought to have first call on them;
4. That if the two legs were taken, a well-directed bomb dropped on either end of the main tunnel would put it completely out of commission, with consequent dislocation of important military and foodstuffs traffic.

Over the coming weeks, numerous letters were sent within the government and local representatives about using the Rendel Street tunnel for use as a storage facility for aircraft engines and parts. It was also noted that the close proximity of the tunnel branch to the docks were susceptible to the constant enemy action (bombing). There were now considerable discussions to use the space under the road tunnel for storage of materials for the war effort, particularly 26,000 tons of oil.

To this aim the government and in particular, the petroleum department discussed the matter with the original company who designed the tunnel Mott Hay and Anderson. Given the costs of the works to allow the storage to be used for oil against the small amount that the tunnels would hold, compared to other storage facilities throughout the country, Mott Hay Anderson stated that the scheme would not be worth entertaining.

The North West was as an important area for the supplies of men, aircraft and materials and from January 1942 American GIs were arriving at Liverpool Docks, American aircraft were being transported to Britain by sea and unloaded in Liverpool before being transported to the nearby Speke, Burton wood and Warton (now BAE Systems) Airports. During WW2 over 1,000 bombers were built at Speke Airport and many of the American Aircraft used by the RAF were shipped to the Pier Head and towed down Allerton Road to the airport before being prepared for active service. 19 million war purpose parts, over 4.7 million troops (Inc. approximately 1.2 million American GI's) passed though both Liverpool and Birkenhead, almost 74,000 aeroplanes and gliders were brought into the port, most of which were sent to Speke Burton wood or Warton and 650,000 tons of tanks and parts

It was anticipated that a surplus could eventually be built up. The planes were to be dismantled and discreetly stored underground somewhere. One proposal was to use 'Central Avenue', the space beneath the main roadway of the Birkenhead (Queensway) tunnel, which was initially intended for trams, to store oil. Various ideas were put forward for this, the most frightening of which was simply to pump the oil straight in until it was full. The City Engineer was asked to investigate the viability of cutting holes in the road surface to make accesses to the lower level for the storage of crates or rubber It was even proposed that the dock branch at the Liverpool end be turned into one long, thin munitions factory for Vickers staff to work without fear of air raids. The fuel task was later to become part of the Pluto Line via Stan low Oil Refinery in Ellesmere Port, which is still operational today. During the war the Pluto Line supplied 1,609 km (1,000 mile) network of pipelines (constructed at night to prevent detection by aerial reconnaissance) to transport fuel from ports including Liverpool and Bristol.

In Europe, the pipelines were extended as the troops moved forward and eventually reached as far as the Rhine.

In January 1945, 305 tonnes (300 long tons) of fuel was pumped to France per day, which increased tenfold to 3,048 tonnes (3,000 long tons) per day in March, and eventually to 4,000 tons (almost 1,000,000 Imperial gallons) per day. In total, over 781 000 m³ (equal to a cube with 92-metre-long sides or over 172 million imperial gallons) of gasoline had been pumped to the Allied forces in Europe by VE day. This provided a crucial supply of fuel.

From the early stages of the war, the River Mersey, along with other strategic ports was seen as a key tactical position and vulnerable to attack from the air and sea. To counter the problems associated with the seaborne threat, the river was regularly swept for mines by the large fleet of Royal Navy minesweepers. This was a relentless and highly dangerous work was carried out by of vessel, including converted former Fleetwood trawlers working out of Wallasey docks.

Ships had to navigate into and out of the port and this task is undertaken by the river pilots. The river pilots are a specialist group of people who know the river and its possible problems well. The men of the Liverpool Pilotage Service had to complete this normally dangerous task in drastically reduced ships lighting, not to mention the lack of shore reference points during air raids to prevent the Germans gaining an advantage with the same reference point. Enemy mines and air raids added to the danger, as did the large number of merchant and naval ships using the river.

By 1945 Liverpool ship owners had lost more than three million tons of shipping, most of which was in the Battle of the Atlantic and the constant supplies coming over from the USA for WW2. This was around a quarter of all British merchant shipping losses (12.5 million tons) during WW2. At first, the Government's policy was to maintain certain exports to help pay for imports, much like today's economy. In late 1940, the high demands of the war virtually strangled Britain's export trade. Almost bankrupt the Government turned to its hand across the sea and asked America for assistance and to bring the supplies across the Atlantic in large convoys.

This approach was the main route of the millions of military equipment and general food supplies that were to prove so vital. This vital supply route became the main target for German bombers to strangle the supply chain and starve Britain into surrender. Known as the longest running conflict of WW2 (it ran from 1939-45) the battle of the Atlantic was of not only paramount importance to Britain and the war effort, but it was to see the Queensway Tunnel as a major part of its organisation. During the conflict 36,200 sailors, 36,000 Merchant Navy personnel, 3,500 Merchant Navy Ships, 175 Warships were lost and most of those lost have no grave but the Sea.

Seeing the tactical advantage and necessity to facilitate the Americans assistance, Winston Churchill ordered a new Command Headquarters was to be set up in Liverpool, and it was to be the main and most central convoy port of the war. In February 1941, Admiral Sir Percy Noble, the C-in-C, Western Approaches, set up his new headquarters at Derby House in the city centre. The building housed a large operations room deep under Derby House, which soon became the nerve centre of Britain's Atlantic campaign.

It was affectionately known as the 'The Citadel', or 'The Dungeon', by the staff that were based there to oversee the day to-day strategy of the battle was directed. The building and command room can be seen today with regular tours of the facility. Liverpool also played a crucial part for the invasions of North Africa, Liberation of Malta, and the D-Day Landings as day and night workers toiled to undertake such things as the tide tables in Bidston observatory in Wirral.

Until mid-1941 only a small force of naval escort ships was based at Liverpool. These consisted of the following:
- Mine sweepers based at Birkenhead for minesweeping and convoy escort work.
- A small number of auxiliary merchant cruisers sailed out of Liverpool to undertake the North Atlantic Patrols. These were, fast and well-armed former liners taken over by the Navy.
- A small group of destroyers based in Gladstone Dock, Bootle.
- A fleet of trawlers from Fleetwood at Wallasey Dock, for early mine sweeping duties

The port of Liverpool, its dock labourers, shipbuilders and ship repairers played a crucial part in ensuring Britain's survival. They cleared ships cargoes, and ensured a good and timely turnaround of the vessels. A lot of this work was interrupted by the constant air raids and which had the dock workers going beyond the call of duty to ensure the supplies were unloaded and stored.

Between April 1940 and April 1941, the main west-coast ports of the UK handled around 60% of Britain's imports, with Liverpool and Manchester taking 31% together. This was more than twice the trade of Glasgow and Greenock at 14.8%, which in itself was a remarkable figure for Liverpool's nearest 'rivals'. This shows the enormity of the task in hand for the port and the tunnels to the war effort, most of the work would go unnoticed but was still vital to the day-to-day running of the river ports.

It was in the early stages of the Second World War, the Queensway Tunnel caught the attention of the Ministry of Supply as well as the Ministry of Transport. Secret correspondence was sent from the wartime ministry and Liverpool City Engineer to see if the tunnel could be utilised for the war effort. William Chamberlain from the Ministry of Transport discussed the matter with the City Engineer who was quick to point out that Liverpool was the busiest port in the empire. This was due to it being the main gateway between the USA and Britain. It also had the statistic that most of its substantial cross-river traffic went through the tunnel. In addition to this, many of the old steamer ferries had been decommissioned shortly after the new road opened.

The possibilities for the utilisation of the tunnel were numerous and there was some discussion as to use part of the tunnel for the war effort and daily transport or perhaps only permitting goods vehicles necessary to the war effort.

One idea was to partition the tunnel via a vertical wall, which would leave two traffic lanes open and the other two for storage. This would allow a long thin space of more than two miles long and 22 feet wide. The more realistic proposal was to close the narrower branch tunnels to New Quay and Rendel Street, or to close the main lines between the branch junctions and the surface, leaving the narrow branches as the only ways in and out.

Investigations into all sorts of applications were carried out and meetings held for quite some time. In the end, however, it was decided not to put the tunnel to any special use. Part of this was down to practicality: storing aeroplanes in a tunnel is not an easy thing to do; if oil were pumped in, it would be very difficult to get out again.

The point was raised that if the Liverpool Dock exit were turned into a factory, there would be advantages to a linear production line of about half a mile in length, but if there was any sort of fire or emergency (always a possibility around explosives) there would be no safe escape route for the hundreds of staff down there. As an example of the tunnels vulnerability, St Nicholas Church (St Nicks) close to the Liverpool dock exit was badly damaged in a bombing raid. A little closer and they may have hit the tunnel dock exit which, if this was used for the war effort may have had a devastating effect on the surrounding area.

It was feared that if certain branches were closed and put to different uses, then one single well-placed bomb could put the whole tunnel out of use by destroying one of the portals. The docks at Liverpool were running at full capacity to support the demands of the war and the tunnel was vital to their operation. There was also the important consideration that the Queensway was being used as an unofficial bomb shelter. Five minutes' walk into the tunnel could put you into a much more secure environment than any of Liverpool's official bomb shelters, and after a time the authorities gave up and permitted this use.

It has to be mentioned that the much forgotten about dock system of Birkenhead, on the other side of the River Mersey, was similar to that of Liverpool and as such it was as important to the war effort and as just an important target for the Germans. Most of the war supplies (aircraft, food, lorries, troops etc.) arriving in Birkenhead would have to have been transported through the Queensway Tunnel to its onward final destination.

It's something of an anti-climax, then, to find that with the exception of moving war supplies through the tunnel, no particular use was found for the Queensway during the war. But it's incredible to think what could have happened under the Mersey if it hadn't proven so difficult to protect. Some of the schemes to assist in protecting the docks along the Mersey (Birkenhead and Liverpool) included setting false fires on the approaches to Liverpool and setting fires to make out that this was the docks on fire. The result of this would be the bombers dropped their bombs on what was effectively a sterile area. In Wirral, Hilbre Island was used to house generators and lights to simulate the docks as it has a similar coastline along West Kirby. One of these generators can be seen today, all be it mostly buried.

Since the Rendel Street branch was closed in the mid-1960s, alternative uses for this section of tunnel have been found once more. This branch has been used for many things over the years from disaster scenarios for the emergency services to sets and locations for major films. There were plans to build flats and an air raid shelter over the Rendel Street entrance and these have recently been discovered by the author during the research for this book and handed to Wirral Archives. Some of the plans are at the end of the chapter to illustrate the development if it had gone ahead.

By the time the German Luftwaffe retreated from British soil in 1942, almost 4,000 people had been killed and 3,500 seriously injured. Merseyside also suffered most lives lost when the bombing stopped mainly due to the close proximity of the residential areas and the docks along with the style of bombing raids adopted at the time. On the Liverpool side, 554 unidentified bodies were interned in a communal grave at Anfield Cemetery. On the Wirral side, there was killed in the Wallasey blitz and are buried in a communal grave in Wallasey Rake Lane Cemetery which notes 324 civilians. Wallasey endured its first bombing attack on August 10th 1940.

The 35th and last raid on Wallasey took place on November 1st 1941. In the raids of Christmas week, 1940, 119 people lost their lives and 91 were seriously injured. March saw one of the most sustained and extensive periods of parachute bombing, during the March blitz, 1941. In three days of continuous bombing, 186 people were killed and 196 seriously injured. Thus, most of the total non-combatant deaths, recorded on the memorial plaque, occurred during these two bombardments.

One of the individual incidents that stand out is at Lancaster Road, Liscard, and Wallasey. On the night of March 12th 1941, a parachute bomb killed 30 people who were killed whilst they were inside an air raid shelter, which received a direct hit.

All the casualties of the war were officially classed as 'Died Due to War Operations'.

The Luftwaffe launched 68 bombing raids on Merseyside between July 1940 and January 1942, which culminated in the infamous 'May Blitz' of 1941, in which several nights of very heavy raids. In all, some 4,000 people were killed and 4,000 seriously injured. Ten thousand homes were completely destroyed and 184,000 damaged with the loss of 4,000 people with many more injured or maimed for life. The largest explosion on Merseyside during the war occurred on the night of Saturday 3 to Sunday 4 May 1941. The ammunition ship 'SS Malakand' blew up in Huskisson Branch Dock Number 2, Liverpool.

The 'Malakand' was fully laden with 1,000 tons of bombs and shells destined for the Middle East and was unable to leave port before an air raid took place over the area. The German aircraft dropped incendiary bombs all around the dock area where the ship was berthed and unfortunately a large barrage balloon which was positioned to protect the area came loose from its moorings and burst into flames on the 'Malakand's deck. Despite several hours of desperate struggle by firemen and crew, under repeated attacks by German aircraft, the ship finally blew up shortly at 3am. Such was the violence of the explosion that a wide area around the vessel was totally devastated. Incredibly, only four people were killed. This story can be heard on a ferry trip from Liverpool to Wirral as part of the history of the river.

The cost of the Battle of the Atlantic was extremely high and the Allied merchant shipping losses (all causes) in the North Atlantic, during 1939–1945:

- 1939: 47,000 Tons of Shipping Lost
- 1940: 349,000 Tons of Shipping Lost
- 1941: 496,000 Tons of Shipping Lost
- 1942: 1,006,000 Tons of Shipping Lost
- 1943: 284,000 Tons of Shipping Lost
- 1944: 31,000 Tons of Shipping Lost
- 1945: 19,000 Tons of Shipping Lost
- Total: 2,232,000 Tons of Shipping Lost

Captain FJ ('Johnny') Walker, CB, DSO and three Bars, was the most famous escort commander to be based at Liverpool during the war. Captain Walker was a specialist in anti-submarine warfare officer who used a somewhat unorthodox method of working. He was held with great respect and affection from the men he served with. Exhausted by the demands of war, Walker died of a stroke in July 1944. He was buried at sea in Liverpool Bay after a funeral service with full naval honours at Liverpool Cathedral. His methods of submarine hunting were to have a lasting effect on the Allied anti-submarine campaign.

After the war, Admiral Horton, C-in-C Western Approaches, considered that victory in the Atlantic was due more to Walker than to any other individual. A statue of Captain Walker has been placed at the Pier Head looking out towards his beloved Liverpool Bay as if forever on duty seeking out the U Boats.

Transportation of Planes through Tunnel from Docks (Liverpool Museum)

German View of Liverpool (Note Bomb Damage Close to Docks) (Wirral Archives)

German View of River Mersey and Docks in Liverpool and Wirral (Wirral Archives)

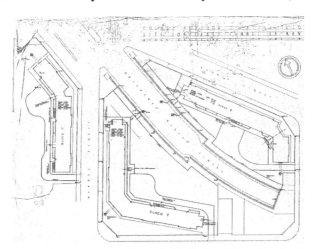

Rendel Street Flats and Air Raid Shelter Plans (Author)

Ground Floor Plan (Author)

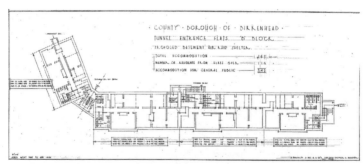

Air Raid Shelter in Basement (Author)

Chapter 18
Queensway Tunnel Approach Scheme, July 1969

Prior to the mid-1960s there were around 3.3 million vehicles using the Queensway Tunnel and the traffic flow into and around the tunnel was now starting to back up in the surrounding areas and town centres. This was also exacerbated by the commercial and retail areas of Birkenhead becoming grid locked due to the queue of traffic waiting to access the tunnel.

In 1964, Mr H. C. Oxburgh drew up plans for a large-scale concept of traffic management to assist in alleviating the congestion suffered in Birkenhead, especially during peak hours. The solution was to start the concrete mixers and start drawing up new flyovers to go at both ends of the tunnel. It was impossible to miss the fact that a whole new tunnel was being built to the north to relieve the Queensway, but that doesn't seem to have stopped a remarkable scheme to demolish huge areas of Birkenhead.

By the early 1960s, it was becoming clear that the Queensway Tunnel was carrying much more traffic than had ever been envisaged when it was first opened. In just thirty years, it was thick with cars and Lorries throughout the day. The problem came in two parts: the capacity of the tunnel, which was set at four traffic lanes; the bottlenecks along its route, and the tollbooths at each entrance.

The capacity issue was slowly being addressed with the planning of the second tunnel and it was generally agreed that the bottlenecks had to be fixed regardless of this, as the Queensway would continue to struggle with its traffic problems even once it was relieved further north. The Kingsway was more for long-distance traffic, and traffic for Birkenhead itself would probably continue to use the old route. For one thing, the junctions inside the tunnel were remodelled. The Liverpool spur became exit-only, allowing traffic from Birkenhead to opt for the Dock exit or the Haymarket exit. The Birkenhead spur to Rendel Street was closed completely for the simple reason that its traffic signals were a liability and caused unjustifiable delays.

The proposed plan would allow different sections of traffic to flow a pre-determined route to the tunnel, local town traffic and local dock traffic to allow them to reach their destination. A system of viaducts and underpasses would separate the traffic and two automatic marshalling areas would accommodate the tunnel traffic. A bill was proposed in Parliament at the beginning of 1965 and it received Royal Ascent on August 5th 1965. It was known as the Birkenhead Corporation (Mersey Tunnel Approaches) Act 1965.

In addition to the powers granted in the 1965 act, the Corporation also implemented the compulsory purchase of land and property under the Chester Street Clearance Area Compulsory Purchase Order 1961. This allowed 179 houses, flats, and private homes, 90 shops, 23 factories workshops and yards, 14 public houses and a variety of other public buildings and land to be acquired.

A large number of those affected by the order, especially local factories were found new premises, many of whom located to land owned by the corporation, and 224 families were found new homes at a cost of £1,000,000 (£15,144,600.00) for the whole operation.

A report in August 1966 gave an outline of the final plan which involved some major changes.

Toll facilities would be streamlined and, because of the higher land prices in Liverpool, all tolls would be collected at a single site at the Birkenhead end of Queensway.

Traffic entering and exiting the tunnel would be segregated and the local road network in Birkenhead town centre would be re-engineered to allow free movement from the major radial routes to the tunnel mouth stopping only for the toll.

To alleviate rush hour congestion towards Liverpool, marshalling areas would be created where traffic queues would be computer-controlled and congestion would not spill onto the surrounding streets.

As with other large projects, even today, there is always an element of ancillary work to be undertaken on site. This would include such items as road and pavement (ground) levels, re-routing and installing gas, water and electrical services, strengthening of other areas close to or at the construction site, re-aligning other roads and such systems to allow the new scheme to be constructed. Areas affected by this scheme were re-aligning or widening and re-routing of Chester Street, Tunnel road, Borough Road, Clock Tower Roundabout (Central Station Roundabout with the listed King Edward VII clock tower and toilets in the middle), and Hinson Street. Any Chester bound traffic is not to pass over the roof of the newly created underpass. This is the second tome the King Edward VII clock had been part of the Queensway works. It was initially moved from its original location next to Central Station around 1929 to the present site opposite the Central Station Hotel to make way for the original Queensway Tunnel.

The traffic marshalling area for the tunnel was designed with a fully automated traffic control access system, which was to provide a free flow of traffic in the centre of Birkenhead and segregating the other traffic for the tunnel from town and docklands traffic. This area, which was previously controlled by the tunnel police during peak traffic times, was fully operational in the early part of 1970. The traffic was directed into two main areas, at the North West and South East of the tunnel entrance. Any vehicle which chose not to use the tunnel or had accessed the area in error would be directed away from this area.

Once in the marshalling area, the traffic would be controlled by feeding the lanes into the manned tollbooths so that vehicles would be fed into the tunnel in an orderly manner especially during peak times. In the early days of the marshalling yard, the driver was requested to turn off their engines and await their turn to proceed to the tollbooth and then onto the tunnel. The system was designed to allow 4500 vehicles per hour through to Liverpool with an option to adjust the systems needs as required if there are few or more vehicles wishing to travel to Liverpool.

This system would not be used today in the Queensway, but is used in the Kingsway during peak times. Vehicles were directed to the correct queue by an overhead traffic light system on a gantry above the access road. Smaller signs were located at other access points and some even had what was described as 'Secret Faces' which would be illuminated when a diversionary route was used to ease the build-up of traffic in an emergency.

Using loop detectors buried in the carriageway (much like todays traffic lights), the control room staff could see a visual chart of what was happening so they could illuminate the secret face signs automatically with co-ordination with the local emergency services. The new system used 27 light masts each 100 feet (30.48 metres) high with a spacing of 240 feet (73.15 metres) apart. The new lighting system was also dual purpose in that it assists in the aesthetics of the tunnel and overall scheme to give a much better lighting scheme compared to standard lighting masts. The masts were constructed from high strength chrome molybdenum seamless steel tubes. These were mounted in a winch mechanism for lowering the lantern to the ground for maintenance and contained four 1000-watt MBF/U colour correction mercury fittings. Some of the other works within this area are the 300 warning and regulatory traffic signs, 46 informative and directional signs, installation of pedestrian crossings and barriers.

On approaching, the motorist could take any lane showing a white arrow at the entrance and join the queue in that lane. The exit from all lanes was blocked by red crosses. The computer system would then allow a small number of lanes to proceed by showing them a white arrow, and at the same time, block the entrance to those lanes with a red cross. Once a lane was empty, it was allowed to fill up again from the back and other lanes were allowed to proceed to the tolls.

There was a simple flaw with this scheme, which was that any regular user would quickly realise that the lanes were released in a sequence and would learn to join whichever lane was next to be released. The less scrupulous could always simply join an open lane and move across into one being released.

The system was installed in 1970 but never really worked properly. The computer system failed at its job quite miserably, and it was rapidly becoming clear that the new Kingsway Tunnel to the north was clearing many of the Queensway's problems. By the time the queuing system was built and active, the epic congestion it was designed for had gone.

Queensway Tunnel Entrance from Birkenhead circa 1970
Note the New Booths and Marshalling Yard Markings (Wirral Archives)

The elevated viaduct designs for the scheme are based on a composite concrete and steel box sections spanning between single columns, which would allow for an aesthetically pleasing design. The design allowed for a shallow deck at a height, which would also allow for a full requirement of vehicle height below the deck. This was achieved by avoiding the use of a transverse supporting beam and allowed for shorter access ramps. The design did incorporate a hidden concrete crosshead beam within the depth of the viaduct beam.

The viaduct girders consist of two or three cell box girders of 12 foot (3065 Metres) and 18 foot (5.48 metres) total width. Due to the length of the spans from 57 foot (17.37 metres) to 110 feet (33.5 metres) and the width of the steel box girders of 12 (3.65 metres) and 18 foot (5.48 metres) were used in the construction of the viaduct. Above the deck, the finish to the viaduct includes the galvanised and painted crash barriers and balustrade, Portland stone parapet, granolithic concrete kerbs, mastic asphalt deck. A 6' (150 mm) diameter drainage system is also incorporated into the viaduct structure.

There are two columns cladded in specially imported Venetian glass mosaic which was applied with a plastic adhesive.

Due to ground conditions and an existing railway bridge over Chester Street, a viaduct was not a viable option for the Chester bound traffic, so a direct graded road was used to link with the road system. The northbound traffic heading towards Woodside and the docks would have to be treated separately via an underpass and viaduct combination. The underpass is a 20' (6.09 metres) wide two lanes single direction road system to take traffic to Woodside and the Docks. Crossing in an almost 90-degree angle on the roof of the underpass, there is a two-lane roadway taking traffic to the other side of Birkenhead.

The design and construction methods of the system had two main areas. Firstly, and throughout the construction, a minimum of one lane in each direction had to be left for normal traffic flow. This requirement meant that the working areas of the underpass and its approach cuttings were severely limited. The second major point was time. In order to complete the required rapid completion of the project, Chester bound tunnel traffic had to be diverted across the roof of the underpass as quickly as possible.

The walls were constructed from a bentonite diaphragm wall method of construction and would therefore enable construction of the permanent reinforced walls of the underpass to be constructed from ground level without the need for an open excavation. Construction of the diaphragm walls was carried out in two phases.

The first phase was completed in September 1967 and the second phase was from February to May 1968. This work consisted of the walls to the southern approach cutting were built together with the roof area which was to be used to transport diverted traffic over the underpass construction area. Traffic was finally allowed to access the new roadway (Underpass roof) in March 1968.

Now that the roof was complete, work could begin on the excavation of the actual underpass using front loaders and lorries to aid the tunnel process of excavation. The diaphragm walls cannot be seen as they are concealed by the asbestos lining, which has been places approximately 6" (150 mm) from the wall to allow natural water seepage to fall into the drainage system. The wall has been finished off with gunite and Portland Stone.

The process of constructing the underpass was as follows:
- A specialist excavator digs out a series of long narrow trenches using a clam operated bucket. To prevent the mud in the trench collapsing, a continuous system of filling it with bentonite mud is used (similar to the construction of the Kingsway Tunnel pile foundations). This congeals onto the earth and thus prevents the cave in of the trench sides and controls the trench width.
- When a suitable depth has been reached (up to 50 feet (15.24 metres) in parts of the construction) a pre-welded cage of reinforced steel is them lowered through the bentonite mud, again, much like the piles on the Wallasey tunnel.
- Concrete is poured into the hole and over the reinforced steel cage, which then displaces the bentonite mud which is then run off into a holding area.
- The diaphragm wall is then complete and once this has cured properly, the roof slab can be cast on the levelled ground surface. This allows the natural ground to form the base of the slab as would otherwise be the case if the roof has to be shuttered and supported whilst it was being poured.
- After traffic, has been diverted over the underpass (on the new roof) the excavation of the actual lower roadway can start
- Finally, the underpass can be completed by installing the, drains and the base of the road were laid along with kerbs, brick dado walling, water mains, and associated hydrants along with an asbestos lining and cornice lighting. Finally, the road was completed and the asphalt road could be laid

The site has no less than four rail systems within close proximity of the construction site. Two of the routes presented no real problem to the contractor and scheme designers, but the old Monks Ferry line had an old arch that would need strengthening as it ran close to the proposed underpass construction area. The Woodside line had been strengthened over 30 years ago as part of the Kings Square Entrance to the Queensway Tunnel, so that was no major problem. The lower level Merseyrail tunnel (still in use today) was of such a depth that it would not pose a problem.

The major problem would be the cut and cover railway tunnel beneath the Haymarket or entrance to the Queensway Tunnel. This tunnel had survived over 80 years of traffic and vibration above, but it was more than likely that the new surge of traffic, not to mention any possible future requirements would be too much for the tunnel. To this aim, it was decided that the only viable option would be to construct a new roof over the railway tunnel. This new roof would be capable of taking the extra traffic and any possible perceived increase in the future.

The new design had to be carried out in several short stages and allow for the railway tunnel and traffic from the Queensway Tunnel to go as smoothly as possible given the works within the overall scheme. In addition to the problems with the new roof, new services (gas, water and electricity) had to be installed / re-routed and connected before the ones in the original route could be cut off. As if this was not enough, the contractor had to phase the works alongside the adjacent roadway viaduct and ramp, which extends across the new roof to bring the Conway Street traffic into the marshalling area. Access had to be given to allow the drains and sewers in the area to be diverted as required. The new roof was formed using pre-stressed concrete beams in an inverted 'T'. Pile capping beams, were supported on a row of 2 feet diameter concrete bored piles approximately 50 feet (15.24 metres) deep and 10 feet (3.048 metres) centres.

The roof has one interesting detail where it crosses the Haymarket railway tunnel in that it was impossible to maintain the 10 feet (3.048 metres) spacing between piles, so a longer and heavier reinforced pile had to be used. The capping beams for these span 60 feet (18.28 metres) across the width of the lower level tunnel and are supported on 1 foot (300mm) diameter bored piles extending 90 feet down to the bedrock. The Monks Ferry brick arch tunnel lies just below the underpass road surface and it was decided that due to the weight of the additional traffic, this would not be able to withstand the additional loads and therefore had to be strengthened. The works were completed by using a reinforced concrete saddle around the outside of the tunnel down to the spring line of the arch (the point at which the curve begins). It seems there may have been an agreement between the tunnels and the local authorities as to who would ultimately be responsible for the costs of the road works and a figure of £5.7 million was paid by the tunnels to the local authorities as recompense for the associated roadwork's.

At the other end of the tunnel, Liverpool had fewer problems to solve. With the congestion generated by the toll booths moved to the other side of the Mersey, its only concern was to get traffic flowing smoothly in and out of the tunnel mouth. By the late 1960s the city already had well-developed plans for an urban motorway scheme and the tunnel mouth had to connect to this. Today the tunnel emerges at a small roundabout close to the main museum and tourist areas. The area is relieved of through traffic by the twin Churchill Way flyovers connecting Islington to Dale Street and Great Crosshall Street. The original 1960s plan was for the tunnel to free-flow onto Byrom Street, which would become part of the main ring road out of Liverpool.

The tunnel Approach Scheme was opened on July 15th 1969 by Alderman Hugh Platt O.B.E., J.P. Chairman of the General Purpose Committee.

His speech was as follows in a special brochure commissioned for the opening commemoration:

"It gives me great pleasure to mark the opening of the great engineering project by recording this message in the brochure specially commissioned to commemorate the occasion. We are now nearing the culmination of one of our efforts to resolve Birkenhead's serious traffic problems, and it highlights the results which can be achieved by hard work and co-operation between public bodies such as the Corporation the Mersey Tunnel Joint Committee and their staffs, the Consulting Engineers and the Contractors. They have all worked together with a single objective – to get the job done on time. It is only just over two years ago, that the then Minister of Transport performed the ceremony to start the work. There is also an historic connection in the 35 years ago, almost to the day (18th July 1934), King George V declared the Mersey Tunnel open to traffic – the same Tunnel as we are using today.

"I am confident that this new complex of roads, viaducts and marshalling areas will improve the flow of traffic through the tunnel and, of no less importance to Birkenhead, will help to relieve congestion in the centre of town for many years to come."

A plaque commemorating the opening of the project listed the various people and companies involved in the project:
- Councillor George William Gill – Mayor
- Ian G Holt – Town Clerk and Chief Executive Officer
- H. C. Oxburgh BSc C Eng MICE FIMun E MTPI – Borough Engineer and Surveyor

- Brain Colquhoun & Partners – Consulting Engineers
- Marples Ridgeway Ltd – Main Contractors

In the initial stages of the construction, a lot of the town had to be razed to the ground to provide the necessary access to and from the tunnel for the flyovers. There was also talk about a possible inner relief route for the areas around Watson Street, Exmouth Street and Rendel Street (Closed dock exit), but this did not materialise.

The main contract started on July 16th 1967 and was completed 28 months later, and the unique automatic traffic control scheme was due for completion in the spring of 1970. All preparation work, drawings and contract documents were completed in 15 months, which for the time and complexity of the contract was seen as a significant achievement. No Computer Aided Designs were available then and everything had to be drawn (and altered) by hand, with the calculations on pen and paper. An emphasis was placed on the final look of the structures as these would effectively partition off two ends of the main town so they had to look right and perform well. The marshalling and approach roads were seen as unique in that it had multiple arrival lanes. This allowed the automatic system to effectively force the vehicles into a more manageable system prior to them entering the tunnel.

The initial entrance point for this fly-over is from the A553 Conway Street and the A552 Borough Road. The Borough Road Flyover is still in existence today, but the Conway Street Fly-Over was demolished in the early 1990s as part of the Wirral Citylands Redevelopment within Birkenhead Town Centre. It was decided that the Conway Street Flyover would be demolished and a roundabout along Argyle Street would be provided for the purpose. This roundabout exists today and has proved to be an easy transition from the flyover.

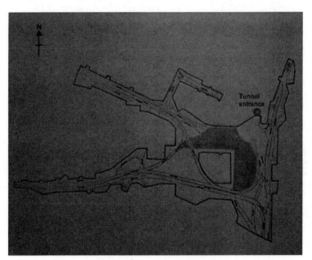

Planned layout of Roads
(Birkenhead Tunnel Approach Scheme 1969)

The Conway Street flyover was positions at the top of the triangle (now the location of the main bus station) leading to Argyle Street and the Queensway Tunnel Entrance / Marshalling Yard.

At the bottom of the picture you will see Borough Road which was the location of the second Flyover and Birkenhead Central station in the immediate bottom right hand corner.

Construction of Flyovers
(Birkenhead Tunnel Approach Scheme 1969)

Construction of Chester Street Underpass (Birkenhead Tunnel Approach Scheme 1969)

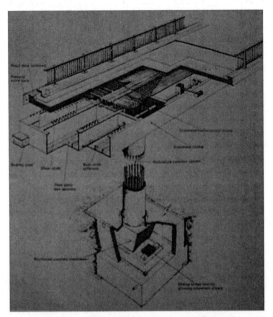

Typical Flyover Construction Detail
(Birkenhead Tunnel Approach Scheme 1969)

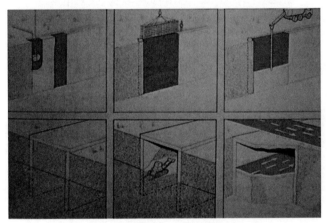

Construction of Underpass
(Birkenhead Tunnel Approach Scheme 1969

Construction of Diaphragm Walling Using the Bentonite Process for the Underpass:

Diagram One – A Specialist Excavator machine digs a narrow trench using a clam grab. Trench is continuously filled with bentonite mud which congeals on the faces of the trench to prevent them from collapsing

Diagram Two – When a suitable trench has been completed to full depth up to 50 feet (15.24Metres) in parts a pre welded cage is lowered through the bentonite mud.

Diagram Three – Concrete is poured into the trench and disperses the bentonite mud which runs into a reservoir

Diagram Four – The diaphragm walls complete the underpass roof is cast on the levelled ground surface

Diagram Five – After traffic has been diverted over the underpass roof, excavation within the underpass takes place.

Diagram Six – The final part of the structure to be completed is the underpass road slab. Finishing works within the underpass then start

A similar construction method will be used on the Kingsway (Wallasey Tunnel) for the pile foundations.

Completed Chester Street Approach
(Birkenhead Library)

Remainder of Old Flyover to Conway Street, Birkenhead (Author)

View down Underpass to Approaching Traffic (Author)

The following Birkenhead Entrance to the Queensway Tunnel Flyover Construction and Demolition Pictures were kindly provided by the Birkenhead Central Library

General Aerial View of the Whole Site for the
New Tunnel Access Routes

BPL 3300   Tunnel Approaches scheme 23/6/1967 – Construction of flyovers

Construction of the Base Units plus
Reinforcement Materials

Independent Supports with Sections of Old Flyover
Awaiting Final Demolition during a weekend of Road Closures

BPL 3137 Demolition of Birkenhead flyover

Section of Flyover from Conway Street that passes by the new Bus
Station and Old Cinema which can be seen to the
right adjacent to the Mersey travel bus.

BPL 3140 Demolition of Birkenhead flyover

Final Dismantling of the Flyover Sections You can see the box section makeup of the roadway in the section above the support.

General Statistics of Scheme

Estimated Value including Purchase of property	£3,000,000 (£48,758,400.00)
Commencement Date of Main Contract	16th March 1967

Viaduct Structure

Total Length of Elevated roadways	3680 feet
Total Weight of Structural Steel	1625 Tons
No of expansion joints	21
Number of columns	37

Column Foundations

No Reinforced Concrete pile caps 4x3 ft 6inch diameter piles	20
No 25x22x5 ft deep reinforced concrete spread footings	17

Ramps

Length of reinforced gravity walls, On mass concrete spread Footings with earth fill in-between	854 feet (260.3 metres)

Length of cellular construction

on mass concrete spread footings 625 feet (190.5 metres)
Length of cellular construction
Situated part of tunnel roof, part of
Piled foundations 210 feet (64.0 metres)
Overall length of reinforced
Concrete gravity walls on piled
Foundations with earth fill in-between 325 feet (99.0 metres)

Underpass

Length of covered section 310 feet (94.48 metres)
Length of open cut approaches 800 feet (243.84 metres)
Average width 2 ft 10 inches (863.6mm)
Thickness of diaphragm walls 1 ft 8 inches (508mm)
2 ft 0 inches (609.6mm)
2 ft 3 inches (685.8)
Depth of diaphragm walling 25feet to 50 feet (7.62–15.24
metres)

Marshalling Areas

Area of South East Marshalling Area 9,000 sq yards (7525.14 m2)
Area of North West Marshalling Area 9,000 sq yards (7525.14 m2)

Haymarket Railway Tunnel

Area of reconstructed roof: 380 ft x 76 feet (115.8 x 23.1 metres)
Beams Spans: 40 feet 4 inches (12.29 metres)
35 feet 7 inches (10.76 metres)
Total number of pre-cast beam units used: 414
No piles 24 inch in diameter: 65
No high mast lights: 27
Power supply: 440V 3 Phase supply
Research and Construction Companies:
Imperial College of Science & Technology London Viaduct crosshead model testing
Model construction: Cammell Lairds
Fabrication of Structural Steel work: Lloyds's Registered Industrial Services
Inspection and testing of precast beams: Sandberg
Lab tests on road materials: Robertson Research Co Ltd
Concrete cube tests (Control series): Liverpool University
Gamma radiography: Gamma Rays Ltd
Preliminary Site Investigation: South Lancs Boring Co Ltd
Preliminary Demolition: Valvine Ltd
Removal of Human Remains: from St Andrews Churchyard W.F.Doyle & Co Ltd
Automatic Traffic Control System: Philips Electrical Ltd and Traffice Signs Division of Willings
Principal Sub Contractors and Suppliers to: Marples Ridgeway Ltd
Demolition: G.T.B. Demolition
Concrete Batching: Plant Pancrete Ltd
Portland Cement: Tunnel Cement Ltd

Reinforced Steel: The Rom River Company
Bored Piling: The Cementation Company
Underpass Diaphragm Walls Found Adila Foundations Limited
Underpass Approach Walls Gunite: The Cementation Company
Underpass Asbestos Lining: Nottingham Suspended Ceiling and Roofers Limited
Fabrication of Structural Steel work: United Steel Structural Company Ltd
Site Erection of Structural Steel work: Carter Horsley (Engineers) Ltd
Site Painting of Structural Steel work: J. D. & S. Tighe Ltd
Supply of Paints to Site: W & j Leigh & Company
Portland Stone: the Stone firms limited
Fixing of Glass Mosaic to Column: John Stubbs (Marble & Quarzite) Ltd
Viaduct Bearings and Drain: Bellows Andre rubber Co Ltd
Viaduct Expansion Joints: P.S.C. Equipment Ltd
Viaduct and Pedestrian Balustrade: Varley & Gulliver Ltd
Polyester resin Adhesives: Stuart B Dickens Ltd
Bituminous and Asphalt Roadworks: Limmer & Trinidad Company Ltd
Cable Duct Laying: B.I.C.C. Company Ltd
Pre-Stressed Precast Concrete Beams: Dow-Mac Concrete Ltd
Pre-Cast Culvert Units: Evercrete Ltd
Granolithic Concrete and Flagging (Labour Only): Campbell & O'Dowd
Supply of Pre-Cast Kerbs and Flags: Premier Artificial Stone Co Ltd
Engineering Class: A Bricks (Staffordshire Blue) Ketley Brick Co Ltd
Road Markings: Cromwell Contractors
High Mast and Conventional Lighting: MANWEB/GEC Street Lighting Ltd
General Traffic Signs: Franco Traffic Signs Ltd
Traffic Crossing Signals: The Plessey Company Ltd

Principle Companies and Local Authority Officers in Charge of Scheme

Birkenhead County Borough Council
Ian G. Holt Town Clerk and Chief Executive Officer
A Thelwell D.P.A. Deputy Clerk
H. G. Lees, BSc (Econ), F.I.M.T.A. Borough Treasurer
H. C. Oxburgh BSc M.I.C.E, F.I.M.T.A, Borough Engineer and Surveyor
D. Butterfield M.I.C.E, M. I. Muni. E, A.M.I.H.E Deputy Borough Engineer and Surveyor
W. T. Mitchell M.I.C.E, M. I. Muni. E, A.M.I.H.E Assistant Borough Engineer
J. H. Leach F. A. I Borough Valuer and Estates

Brian Colquhoun and Partners

P. Rushton BSc, M.I.C.E, A.M.I.H.E Engineer in Charge
J.B. Shaw BSc, M.I.C.E, M.ASCE. Co-Ordinating Engineer
A.G. Lance F.I. Struc. E Chief Structural Engineer
G. R. Newman BSc (Eng), M.I.C.E, M.I.Struc.E Chief Structural Engineer
S. R. Gray BSc, A.C.G.I, M.I.C.E, Senior Bridge Engineer
A. H. Shearling A.R.I.B.A Chief Architect
E. P. Windett F.I.E.E, Dip M.I.E.S Chief Electrical Engineer
R. S. Colquhoun M.A, M.I.C.E, A.M.I.H.E, A.M.ASCE Resident Engineer

J. Leech BSc Tech, M.I.C.E.
J. A Forster BSc, Deputy Resident Engineer

Marples Ridgeway Limited

A Sunderland PhD, BSc, F.I.C.E Director in Charge
D. E Shepherd BSc, F.I.C.E.
F. D. Unwin B.A, B.A.I, M.I.C.E Contracts Manager
G. H. Greenbank B. Eng, M.I.C.E Site Agent
D. D. D. Lloyd BSc, A.C.G.I Site Construction manager

Other Authorities and Statutory Bodies Involved in the Project

Cheshire Constabulary
British Railways
Central Electricity Generating Board
Merseyside & North Wales Electricity Board
General Post Office (Royal Mail)
North Western Gas Board
Wirral Water Board
Appointees Involved in Tunnel Approaches Scheme
Members of the General Purpose Committee
Alderman High Platt O.B.E. J.P. Chairman
Alderman J. W. Oates Deputy Chairman
Alderman G. F. Davies J. P
Alderman J. Furness J. P.
Alderman C. S. McRonald
Alderman R. Pilkington
Alderman J. H. Roberts J. P.
Councillor D.T.P Evans J. P
Councillor D. A. Fletcher J. P.
Councillor J. R. A. Flynn
Councillor Miss E. M. Keegan
Councillor C. Pyke
Councillor Mrs P. A. Roberts
Town Clerk and Executive Officer: Ian G. Holt
Deputy Town Clerk: A. Thelwell D. P A.
Mersey Tunnel Joint Committee
Appointed by Liverpool City Council
Alderman H. MacDonald Steward Chairman
Alderman C. Cowlin
Alderman B. Crooks
Alderman A. B. Collins M.B.E.
Alderman J. Keegan
Alderman H. Lees
Alderman A. W. Lowe O.B.E
Alderman W. McKeown
Alderman J. Morgan
Alderman A. Morrow

Alderman J. S. Ross
Alderman W. H. Sefton
Councillor R. F. Henderson
Councillor N. A. Pannell
Appointed by Birkenhead County Borough Council
Alderman H. Platt O.B.E Chairman
Alderman J. Furness
Alderman J. W. Oates
Alderman R. Pilkington
Alderman J. H. Roberts
Councillor D. A.Fletcher
Councillor A. Lindfield
Appointed by Wallasey County Borough Council
Alderman J. P Ashton M.C.T.D
Alderman R. J. H. Collinson B.C.L, M.A
Alderman H. T. K. Morris F.C.A
Councillor C. Allman
Councillor T. F. Cavanagh A.D.I
Councillor C. J. Smith B.A.
Councillor C.G. Tompkins M Inst. M
Clerk and General Manager
G. T. Jones F.I.C.E, F.I. Mun. E

Chapter 19
Emergency Exits

Compared to open roads and motorways, the risk of accidents in road tunnels is minor due to the lack of peripheral additional occurrences such as pedestrians walking out or vehicles pulling out of driveways and side roads.

Statistics show that fewer accidents happen in tunnels than on open roads, which is primarily due to the minimum effect of weather conditions, to speed limits, steady lighting conditions, as well as the low number of junctions/links in tunnels.

But, even small accidents are difficult to manage in tunnels; particularly rescue personnel (ambulance personnel, fire brigade, etc.) have very restricted access. Accidents resulting in fire can lead to a disaster and general access problems, as events in Montblanc Tunnel and Tauern Tunnel demonstrated.

Concerns were raised about how to improve the capacity for emergency evacuation in case of an incident in one of the Mersey Tunnels. The original configuration of the tunnel links between the tunnels was not entirely suitable as an escape provision. Discussions between Mersey Tunnels and Merseyside Fire Brigade identified the need for additional links between the road tunnels to satisfy their evacuation strategy in case of an incident within one of the road tunnels. It was considered that provision of three additional cross passages within the central section of the tunnels, located between the two existing ventilation shafts satisfied the objectives of the evacuation strategy

Mersey Tunnels commissioned Mott MacDonald to investigate and design the passages in the Kingsway as well as the emergency refuges in the Queensway Tunnel. It was decided that seven emergency refuges (A–G) were to be used in the Queensway Tunnel and three routes (nicknamed Tom, Dick and Harry by Mersey Tunnels) were to be used in the Kingsway Tunnel. The design of the cross passages would have to comply with the Highways Agency document BD 78/99 Design of Road Tunnels August 1999.

This document sets out the requirement for cross passage spacing for passenger routes through fire doors, which are positioned in the tunnel walls along with many other factors affecting tunnel design. These fire doors are to be placed at 100metre intervals and be permanently illuminated via signage and have emergency telephones and other facilities such as emergency water supplies. The Kingsway Tunnels were to be positioned equally along the tunnel to miss the geological faults and be located between the existing ventilation shafts. This decision had the new passages at 325 metre intervals rather than the 100 required on the Highways Agency document. A decision was made that given the tunnels location and the other aspects of the design, the scheme was given the green light.

The design team for this tunnel were:
- Peter Arch Mersey Tunnels Engineering Manager
- Tony Rock Mott McDonald Project Director, (The original tunnel design company)
- Frank Rogers AMCO Contracts Director

Construction of the Kingsway Tunnel cross passages commenced in January 2002. The Contract programme set by Mersey Tunnels included for an 'out-of-tunnel' completion date of July 31 2002 and overall completion date on August 11 2002.

Constraints imposed on AMCO Donelon during the construction works included:
- Traffic management, requiring close liaison with Mersey Tunnels Police (the works being carried out during night time possessions available in only one tunnel at a time)
- The presence of tunnel services and requirement for service diversions
- Control of silt in the invert of the Kingsway Tunnels to minimise disruption to Mersey Tunnels sump pumps

Excavation Procedures

Before the excavation of the cross passages, AMCO Donelon relocated the numerous services within the tunnel sections so they now sat within designated DADO trunking before removing sections of the tunnel walkway.

At the inverts of the tunnel, blockwork bunds were constructed to resist the flow of ground water and silt for the operation, along with rock dowels being inserted to support the existing roadway and above the openings. These were fabricated, welded to prevent movement of the existing tunnel segments. Remote access movement devices were also installed to check on the possibility of any background movement of the tunnel segments.

Construction of Cross Passages 1–3 started in April 2002, with the initial phases of excavation and tunnel ring construction starting in the south tunnel. Break through to the north tunnel occurred on June 23rd 2003 at CP 1 and June 30th on CP 2 and 3.

On completion of the excavation and breakthrough, the contractor started the construction of the cast concrete linings through the passages to the road tunnels. Initially this work was completed in the south tunnel, under tunnel closures and was completed by July 29th 2002 when work moved to the north tunnel

During the contract, Mersey Tunnels identified the need for two new under deck Invert Access Passages between the two road tunnels to be located at each end of the Kingsway Tunnel between the portals and the ventilation shafts. These passages were to assist Mersey Tunnels in their maintenance operations as the current configuration of the tunnels only allows access below the deck level at the mid-river section and at each portal.

Although, works did run over the original programme, it was recognised by Mersey Tunnels that this was necessary given the increased scope of work and achieved with minimal disturbance to the operation of the tunnels. The completed work provides additional emergency facilities for Mersey Tunnels operations and improved under-deck access for their maintenance staff. Amendments to the originally planned refurbishment works of the of the mid-river lay-by will result in reduced maintenance and increased life expectancy for this structure

In April 2004, work started on each of the seven new Queensway Tunnel refuges, which had a capacity of 180 people and was the start of a £9,000,000 project to bring the tunnels into line with the highest European Safety Standards. Each refuge is 21 metres (69 feet) long and 3 metres (9.8 feet) wide, accessible from the main tunnel walls.

The refuges have fire-resistant walls, ramps for wheelchair access, a good supply of bottled water, toilet facilities and a direct video link to the Mersey Tunnel Police Control Room. All seven refuges are linked by a walkway below the road surface with exists at both the Liverpool and Birkenhead exists of the tunnel.

To assist in the design of the refuges and cross passages, Mott MacDonald and Mersey Tunnels requested the assistance of RADAR (Royal Association for Disability and Rehabilitation) who played a key role in the practical evolution of a full-scale model and advice on solutions to the problem. The refuges were to enhance the safety requirements as had been outlined in such tragedies as the Mount Blanc it would become necessary to leave the tunnel under a safe controlled manner.

The existing shafts were to be used for new exits and the George's Dock exit was to utilise the second shaft, which had been sunk on the site of George's Dock to facilitate the task of excavation. At Morpeth Dock, which had only one shaft on this side of the river, it was decided that this should be utilised for the same purpose.

Passengers would exit from the tunnel to the shafts via a newly constructed staircase. Originally, the entrances to this escape routes were distinguishable from a considerable distance away, depending where you were in the tunnel at the time. The use of illuminated sign with the title 'Emergency Exit' and a red arrow pointing in the direction of the exit were the only clues you had, and in thick black smoke, these would not have proved efficient enough. On the opposite wall of the tunnel a similar illuminated sign had been erected to ensure the exits was clear.

The George's Dock exit leads along a passage and a short final staircase, to the roadway close to the ventilation station. The Morpeth Dock exit is situated in Shore Road, Birkenhead, close to the old Canning Street Railway Building and connected by a passage and staircase leading from the shaft. The passage from the Morpeth Dock shaft to the final staircase had to be constructed in circular form, with a cast-iron lining, to protect it from water in the surrounding strata; otherwise, all these passages and staircases have been constructed in the ordinary way, and are lined with concrete.

The treads of the staircases in the passages and on the landings, were finished in granolithic, which was laid by Adams Brothers (Liverpool) Ltd who also covered the whole width of the concrete roadway in the tunnels with granolithic prior to the laying of the cast-iron setts.

The new escape refuges originally started design concepts which included Cross passages, to the adjacent rail tunnels, a central avenue connection, aircraft chutes from carriageway to invert levels or alternatively from the walkways. All proposals would have to meet strict criteria on the safety of the tunnel users, including any disabled users.

Design Options

In excess of 10 options were initially considered and reviewed by Mersey Tunnels and Mott MacDonald.
   The options divide broadly under 4 headings:
   1. Escape by cross passage to the existing railway tunnel or to a new parallel escape facility

2. Enclosed and pressurised side-walkway
3. Escape to a refuge in the tunnel invert from the carriageway using chute technology as in aircraft emergency evacuation
4. Escape to refuge in the tunnel invert using ramp access in the side-walkway

The selection criteria were agreed and their importance weighted with Mersey Tunnels as follows, with weights specified below as a percentage.

As you will see, the capital cost to the tunnels would be only 1% and the majority of the cost would be to ensure the safety of the people using the tunnel system.

- 27% – Short travel distance to safe refuge
- 15% – Reversible action, e.g. avoid sliding down chute as return is not possible
- 13% – Disabled person friendly, e.g. step free access, gentle gradient, short distance to safe refuge
- 13% – Support self-rescue, e.g. ensure ability to reach safety without assistance even if necessary in the case of the disabled
- 12% – Non-claustrophobic, e.g. avoidance of long narrow passages, provision of spacious well lighted environment, provision of reassurance from authority by providing two-way oral and visual communication
- 9% – Short travel distance to open sky
- 6.1% – Proven technology, e.g. avoid requirement for complex systems to provide automatic traffic management to facilitate escape
- 5% – Operations impact, e.g. minimise impact on the travelling public in normal operation both during construction and thereafter
- 4% – Value added to adjacent infrastructure, e.g. cross passage to parallel railway tunnel affords enhanced escape facility to the railway tunnel
- 1% – Capital cost

Using the above criteria and comparing all options, option 4 was seen as the preferred option. The concept assumes that motorists in an emergency escape unaided through a fire within an enclosure such as a tunnel enclosure at road level and along a descending ramp to a safe refuge beneath the carriageway.

The design criteria of emergency escape facilities for road tunnels are currently unknown compared to say a new airport design. When a new airport has been designed, various computer modelling will be able to give the designer a specific idea on bottlenecks or possible problem areas for an evacuation. With tunnels and in particular the Mersey Tunnels, this would not be the case. It was agreed that site-specific criteria should be used for the Escape Refuges

The criteria addressed the following: time to close tunnel to incoming traffic 300 sec, based on 60 sec for incident alert, + 120 sec to instruct closure, + 120 sec to implement closure, passenger escape load; number and distribution of cars and busses and vehicle occupancy level leading to 0.6 persons/m length of tunnel i.e. 0.3 persons/m/lane, based on vehicle speed of 14m/sec (30 mph), 1500 cars/lane/hr at 1.2 persons/car, + 30 light goods vehicles/hr at 1 person/vehicle, + 30 coaches/hr at 40 persons/coach stopped vehicle density, cars at 5m/vehicle, light goods vehicle and coach at 10 m/vehicle response times, coach exit times, walking speeds, refuge entry speeds, based on: people in cars; alert at Control Room (CR) 60 sec, further 120 sec to instruct people to leave cars, + further 60 sec to leave car and walking speed of 1m/sec.

People in coaches as cars except; 30 sec to obey CR instruction to leave, passengers alight coach at 2sec intervals, exit through emergency escape door 1 persons/sec.

Time to reach place of safety, 1st passenger from all vehicles to have possibility to reach place of safety within 360sec. All persons to reach place of safety within 600sec.

A calculation based on the above criteria assuming a coach blocking a refuge, confirms that all people reach a place of safety within 600 seconds and car passengers reach the place of safety within 360 seconds, if refuges are provided at 200m centres.

A specific requirement within the design had to be given to any disabled passengers who may be in the tunnel at the time of the emergency and this would fall into three specific categories and these would be fundamental in determining the final layout and design.

Wheelchair / Mobility Impaired access

As the current walkway is currently above the roadway, a change in height of around 1 metre, a method of gaining the height for the wheelchair would have to be found to enable the wheelchair user to access the cross passage. A survey was carried out as to this problem and the exact level of the corresponding road in the adjacent tunnel.

Passenger Escape Width Requirements

There are no specific requirements this other than the building regulations used by any architect or engineer to assist in a design. A width of 1.8 metres was chosen as the desired requirement with a clear height of 2 metres. This would allow three people to pass side by side or a wheelchair bound person pass a person along the passageway. All this resulted in a passage width of 3.35 metres in diameter cross passage lining with a 2.4 metre two ring opening with double fire doors to the tunnels.

Evacuation Method

Traditionally, hand excavation, which would result in white finger (Hand arm Vibration Syndrome), would not only be long and expensive, it would have a detrimental effect on the workers.

It was decided to utilise a small excavator type machine (Brokk Robotc Breaker) to extract the base material from the cross passages and perform the bulk of the excavation. The ground was carefully supported and monitored throughout the construction of the cross passages, and great care was taken to look at the historic information for the tunnels which was taken during their initial construction.

As you can see, the design of a new refuge is not just a simple task of drawing it up, tender the works and issue the works contact, and get on with its construction. There is a considerable amount of unseen work and thought that goes into a design of any type, particularly that such as the Mersey Tunnels.

The construction has brought the tunnel into line with the highest European standards. The road access hatchways where updated to form a 2 x 2 metre hatch to allow the materials to be lowered through the road deck onto a railway system used to transport the materials, plant and equipment to the required locations which allowed full access to the lower central avenue area. This area was originally designated to be used for trams and busses through the tunnels under the road deck. The majority of the work was carried out by means of numerous night time lane closures with live traffic passing in two open lanes. This opening was only available for two weeks in every six which made the whole project a complex planning and logistical exercise.

At the time Councillor Mark Dowd, Chair of Merseytravel, said:
"Safety is of paramount importance and while new European legislation does not strictly apply to the Mersey Tunnels we were totally committed to improving escape provisions to meet the highest modern safety standards."

Neil Scales, Chief Executive and Director General of Merseytravel, said:

"The Mersey Tunnels have an extremely good safety record and we are spending some £32 million upgrading and improving them to ensure we maintain this record."

Part of the £32 million on safety improvements included:
- Construction of seven new 180 person emergency refuges in the Queensway tunnel which will include accessible ramps, new fire protection enclosures, complete with visual and audio links to the emergency services.
- Complete renewal of water pipes to the tunnels fire hydrants
- New gantry mounted message and lane closure signs to assist the driver with information in an emergency
- Enhanced CCTV coverage within the tunnel which will include the latest incident and smoke detection technology
- A programme of modernisation on the tunnels ventilation fans which are almost 80 years' old

Engineering consultant Mott MacDonald, who designed the Queensway Tunnel in the 1920s, was commissioned to review and develop an escape concept. The main contractor on the project, which has also included the upgrading of the existing escapes to the surface at either side of the river, was Amco Donelon who was also extensively involved in the development of the final design for the project, providing alternative design proposals and constructability input to the design solutions.

The refuges have been built under the roadway in a programme that saw the tunnel completely closed only twice during construction of the refuges. Both closures were over weekend periods and the vast majority of the work at road level was carried out on Monday to Friday nights utilising routine maintenance closures of two lanes of the four-lane tunnel.

In the event of an emergency, motorists will be directed to the closest refuge via a public-address system and by directional noise beacons and flashing arrows. Once inside, the refuges motorists can communicate via a two-way audio-visual link with the Mersey Tunnels police control room. The refuges can accommodate people until emergency services guide them either back to their vehicles or along the tunnel under the road deck to shaft exits at either the Liverpool or Birkenhead ends.

Work on the refuges full-scale mock-up designs started in May 2003 to determine the design and how to meet the optimum design requirements. Design of the cross passages focussed on areas that would determine any spatial requirements within the confines of the existing tunnels for Wheelchair/mobility impaired access and passenger escape requirements. With this in mind, the contractor would then be able to ascertain which construction method to use.

At the Birkenhead end of the tunnel (Shore Road) a 45-metre-long, three-metre diameter tunnel was bored through sandstone to provide additional access to the existing emergency staircase leading to the surface at Shore Road.

Traditionally, hand excavation techniques involving miners using hand held compressed air tools would have been employed. Guidelines introduced at the time to prevent Hand Arm Vibration Syndrome (white finger) prevented this type of construction unless necessary. A small excavator/concrete breaker was be used to perform the bulk of the excavation.

As with all tunnel excavations, careful monitoring was undertaken at all times and any necessary adjustments or structural supports installed as required. Further improvements included gantries throughout the tunnel to provide up to the minute driver information and emergency instructions via variable message displays.

Fire Exit signs were placed every 50 metres in the tunnel, showing the nearest means of escape or refuge areas. Other bespoke signs such as 'Mind Your Head', 'Do Not Return To Vehicle' and 'SOS Firebox' were also installed.

In the event of a total loss of power, the photo luminescent signs will continue to deliver the safety measure extremely effectively. Emergency signage is designed such as to be instantly recognisable and comprehensible to convey useful information to both the motorist and the Tunnel Operator.

Criteria for the Signage Was as Follows

Readily visible and conspicuous under difficult conditions i.e. be self-illuminated where practicable, be suitably located e.g. at low level below stratified smoke level, be distinct from normal signage e.g. thermo-luminescent, flashing at frequent intervals etc.

Intelligible, i.e. be of sufficient size to be read from a distance, use standard pictograms in preference to a message, use English only as language

Unambiguously located and or labelled i.e. state clearly the location in the tunnel e.g. distance from portal, labelled as door 3 of refuge 1 or camera 2 of refuge 4 etc.

Durable and robust, signage will require having a long life under normal operation in an aggressive environment and resisting possibly high temperature in an emergency 200 C for 1 hr.

Reassuring e.g. to motorists inside a Safety Refuge by providing clear information on; the location and onward connecting routes, the recommended next action, anticipated action from Tunnel Operator such as to await announcement over PA and similar.

During an emergency, communication at carriageway level is one-way from the Tunnel Authority to motorist and principally of an audible nature. These communications are required to; be clearly audible in the tunnel environment alert to danger, ensure if required that motorists on instruction leave their vehicles as quickly as possible and are aware of the imminent danger given that in some instances the danger may be unseen e.g. hidden by a row of high sided vehicles, danger from radioactive material or invisible gas inform by providing sufficient data to facilitate self-rescue in the most appropriate direction and help to reassure motorists and combat panic and dis-orientation

Having reached a Safe Refuge two-way visual and oral communication will be possible between motorists and Tunnel Operator personnel. This communication is effected using monitors, cameras microphones and loudspeakers.

In relation to the Mersey Tunnels, and as part of the 2005 (EUROTAP) Inspections the AA made the following comments on the Kingsway Tunnel, which was, one of 49 tunnels inspected in 2005, was given a 'good' safety rating. The Kingsway Tunnel was one of the busiest in the 2005 survey – about 45,000 vehicles use it every day, of which 7 per cent are heavy goods vehicles. The tunnel has consistently had a 'good' rating since the inspections started in 2000, and a programme of continuing improvement, described later, should see it moving into the top 'very good' category in the near future.

The inspector noted the following positive points about the tunnel:
- Around the clock manning with trained personnel;
- Well-marked cross-connector escape route passages between the two bores;
- Escort vehicles for hazardous loads;
- Special fire-resistant lining to tunnels;
- Automatic monitoring and control of ventilation systems; and
- A special ventilation programme to clear smoke away from standing traffic.

The inspector also noted some issues that scored less well:
- No automatic traffic (problem) detection system, and no video recording of traffic incidents;
- Emergency phones not insulated against traffic noise;
- Emergency walkways very difficult to access and use; and
- No automatic fire alarm system.

British tunnels have so far had a good safety record through good management, the installation of safety systems to the latest European standards, and an information programme to make people aware of what they should do in a tunnel emergency.

The actual levels and positions of the two Mersey Tunnels were surveyed as a part of a comprehensive investigation before commencement of the new refuge and other works.

Below Road Deck of Queensway Tunnel (Author)

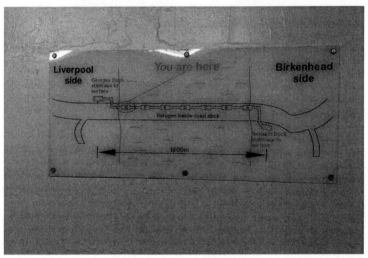

Sign Noting the Location of Refuge within the tunnel Complex (Author)

Reassuring sign within Escape Areas (Author)

One of the Escape Routes (Mersey Tunnels)

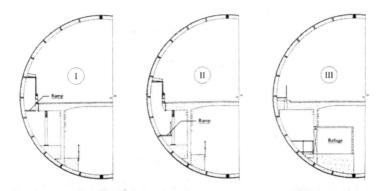

Figure 4: *Cross-section through ramp access*
Ramp Access Details (Mott McDonald)

Shore Road Tunnel Emergency Exit (Author)

Chapter 20
Tunnellers Memorials

Accidents

The table below covers the first four years of the work shows that the frequency rate of accidents was rather high. For the whole period covered by the table, there were 0.10 fatal accidents and 6.23 on-fatal accidents per 100 000 hours worked.

Year	Accidents		Ave No Workers Hours Worked	Hours Worked
	Fatal	Non-Fatal		
1967	0	6	145	281,977
1968	0	72	429	1,177,356
1969	4	62	344	1,128,480
1970	0	102	456	1,289,269

There is a plaque to each of the tunnels (shown below) detailing those that had paid the price during the construction of the tunnels. The Queensway plaque can be found on the St Georges Dock building. The Wallasey (Kingsway) plaque can be found along the Wallasey promenade ventilation shaft, close the Seacombe ferry terminal.

I have managed to find two accounts of how the men died during the construction of the tunnel and these are noted below.

Queensway Tunnel (Birkenhead) Memorial

Memorial to those lost during the construction of the
Queensway Tunnel Construction
The memorial is within the
wall of the George Dock Building

THIS MEMORIAL WAS ERECTED AS PART OF
THE QUEENSWAY TUNNEL DIAMOND JUBILEE CELEBRATIONS
IN REMEMBRANCE OF THOSE WHO DIED DURING
IT'S CONSTRUCTION
WORK ON THE TUNNEL WAS COMMENCED
ON 16TH DECEMBER 1925 AT THIS SITE
BY
HER ROYAL HIGHNESS PRINCESS MARY
AND THIS GRETA FEAT OF ENGINEERING
WAS OPENED AT OLD HAYMARKET ON
18TH JULY 1934
BY
HIS MAJESTY KING GEORGE V

JOSEPH BONNER
26th July 1926, Aged 26

JOHN JOSEPH MCNULTY
17th September 1928, Aged 29

John McNulty lived at 337 Price Street Birkenhead and at his inquest, the Coroner Mr Joseph Roberts, along with a jury heard the account of Mr McNulty's accident and subsequent injuries. Mr John Blakely (miner) of 8 Planet Street rock ferry said that at 5.45 am on Saturday 12th, he and six others were working on scaffolding near the mid tunnel.

Staging was 18 (18 feet) high and suspended on bearers and planks. Mr McNulty was going to put two bolts in a plate on the side of the tunnel and stood on a rock ledge and the side of the tunnel to move a plank. The plank was wet and greasy and the Mr McNulty slipped trying to maintain his hold and fell to the floor. Witnesses went to his aid and Mr McNulty was taken to hospital. Witnesses said the accident happened 2–3 minutes before the end of the duty.

Mr McNulty was admitted to the general hospital at 7 am with a broken right arm, cuts to his face, which turned septic, and bronchial pneumonia, which resulted in the death of Mr McNulty at 4:45 p.m. on the 17th, and three days later, Mr McNulty died.

JAMES HERBERT BROWN
29th November 1928, Aged 18

During blasting for the ventilation shafts the workers were warned to retreat to particular point that was deemed safe, unfortunately what the engineers were not aware of was a large void above the area. This void was supposed to have been created by attempts to undermine the outer wall of Liverpool castle during a conflict. This area collapsed as a result of the blast, sadly dropping numerous tons of rock onto some of the men in the tunnel.

As a result of this tragedy and perhaps the young age of one of the victims, a labourer named James Herbert Brown, aged 18, was killed. His colleagues requested that they be allowed to attend his funeral but the directors refused, and as a mark of respect, the whole workforce took the day off without pay and attended his funeral. It has been said that it was a sight to behold as scores of men escorted his coffin down Smithdown Road to Toxteth cemetery. It was James's relative, also called James after him, who started to cause a fuss about the lack of a memorial and was victorious when a plaque naming all those who lost their lives was erected on the ventilation shaft at Mann Island during the tunnels jubilee celebrations.

JOHN WILLIAM BLAKELEY
7th July 1929, Aged 34

JOHN MCNICHOLAS
13th September 1929, Aged 55

HENRY FRANCIS GARRETT DE MOUL
28th September 1929, Aged 25

JOSEPH COLLEGE
11th December 1929, Aged 62

JAMES MICHAEL WILMOTT
27th December 1929, Aged 42

JOHN CARBERRY
24th March 1930, Aged 26

ALBERT WHITE
27th November 1930, Aged 42

ALFRED PITMAN DUKE
16th July 1931, Aged 45

HENRY DETITH
15th September 1931, Aged 24

FREDERICK JOSEPH DURR
29th September 1931, Aged 33

THOMAS ARTHUR BECKINGHAM
16th May 1933, Aged 57

JOHN CARR
14th July 1933, Aged 23

JAMES GREEN (Bir/182/53)
14th November 1933, Aged 22

DONALD LESTER
15th September 1934, Aged 24

Kingsway Tunnel (Wallasey) Memorial

Memorial Stone on Wallasey Ventilation
Station Kingsway Tunnel Workers

Who Died During its Construction (Author?)

Inscription at bottom reads
THIS MEMORIAL WAS ERECTED IN 1997 IN REMEMBRANCE OF THOSE WHO DIED DURING THE CONSTRUCTION OF THE KINGSWAY TUNNEL BETWEEN 1966 AND 1974

FRANCES ADDERLEY
15th June 1970, Aged 59

RONALD (ROY) CARRY
12th June 1969, Aged 26

BERNARD GLENN DENNESS
1st November 1969, Aged 25

CHARLES KEEGIN
1st November 1969, Aged 28

JOHN LATHAM
23rd October 1971, Aged 33

LEONARD MILLS
8th March 1969, Aged 47

JOZSEF NYARI
23rd October 1971, Aged 32

GERALD RANDELS
3rd March 1971, Aged 38

DANIEL JOSEPH SWEENEY
1st November 1969, Aged 27

Chapter 21
Update on Companies

Update on Companies involved in Queensway / Kingsway and we have tried to update most of the company particulars here. In some cases, the company had ceased trading for a variety of reasons or they have been taken over by other companies and formed into large multinational groups. There are a few small companies who have proved difficult to find, possibly due to their small size and local work.

Nuttall (Now BAM Nuttall)

In 1865, at the age of 21, James Nuttall set up a business as a small contractor for Road and Sewer work, with an office and yard in Burlington street Manchester.

In 1902, James Nuttall took his two sons Edmund and James into partnership, trading under the name Edmund Nuttall & Company.

In 1904, James Nuttall died; the business was carried on and rapidly developed by his son Edmund, who had a natural flair for assessing the price of a job and the right way to carry it out. Nuttall's involvement in the Queensway Tunnel was not their first large scale project in Liverpool. In 1908, they built the Royal Liver Building in Liverpool the first reinforced concrete 'Sky-scraper' built in Britain, 300 feet high to the top of the Liver Birds.

In 1925, Edmund Nuttall secured the contract for the pilot headings of the Mersey Tunnel and eventually carried out the greater part of the civil engineering work. The contract for the Mersey Tunnel lasted six years and in total amounted to nearly £4 million in value. This proved to be a turning point in the history of the company whereby it grew up from a mid-sized firm operating mainly in the North of England to its present status as one of the major civil engineering contractors in the United Kingdom and internationally.

In 1931, work began on King George V Dry Dock in Southampton. At the time of construction, it was the largest Graving Dock, or dry dock, in the world. The dock was opened by King George V and Queen Mary in July 1933.

1945: Edmund Nuttall, Sons & Co Ltd began construction of Ipswich Power Station which was one of the first to be completed after the war.

1946: the company began construction of the Claerwen Dam, the largest of a number of dams built to provide water to the growing city of Birmingham. The Claerwen dam was commissioned by Queen Elizabeth II on 23rd October 1952, one of her first royal engagements as Queen.

Nuttall have also constructed Channel Tunnel, Dartford Tunnel and in 1962 the Tyne Tunnel. This was all flowed by the second Mersey Tunnel, which was started in 1697.

Many changes to the Company structure took place at the end of the last century and the beginning of this. In 1978, the company was acquired by Hollandsche Beton Groep (later HBG.) In 1979, Nuttall acquired Mears followed by Hynes Construction in 1992, John Martin Construction in 1999 and Finchpalm Ltd in 2000. In 2002, HBG was acquired by Royal BAM Group.

Mott (Now Known As Mott McDonald)

Mott, Hay and Anderson, and Sir M MacDonald & Partners is now known as Mott MacDonald was formed in 1989 through the merger of the companies.

Mott MacDonald continued to expand their international presence in 2013 with the purchase of Habtech Engenharia Ambiental and the purchase of the Brazilian-based environmental consultancy.

It also purchased P.D. Naidoo & Associates nine days after the purchase of Habtech. P.D. Naidoo was a consultancy based in South Africa. The acquisitions added to the company's previous year purchases of Canadian consultancy Engineering Northwest, oil and gas firm Procyon, and the oil and gas operations of Mouchel. In 2014, Mott MacDonald acquired AWT, a specialist water technology and consulting company based in New Zealand and Australia.

Mcalpine (Sir Alfred Mcalpine)

Alfred McAlpine was one of the sons of 'Concrete' Bob McAlpine and he ran the operations of Sir Robert McAlpine in the north west of England. McAlpine's status as a civil engineer was enhanced during the 1960s by its participation in the motorway building programme and the company became one of the country's leading civil engineers. There had been some limited diversification, including the purchase of Penrhyn Quarry, the country's largest slate works.

As the civil engineering market declined in the 1970s, McAlpine sought to diversify further into private housebuilding. Acquisitions included Price Brothers in 1978; Frank Sanderson's Finlas in 1982; and Canberra in 1988. Investments had also been made in the US housing industry. By the end of the 1980s, private housebuilding was contributing the major part of group profits.

Under new management, there was further concentration on private housebuilding, including the acquisition of Raine Industries. By the late 1990s, McAlpine was building over 4,000 houses a year and was one of the industry's top ten. However, there was increasing speculation over the future of the Company and, in 2001, it sold its housebuilding operations to George Wimpey. In 2001, it acquired Kennedy Utility Management for £52m. In 2002, it acquired Stiell, a facilities management and information technology network systems business, for £85m. In February 2008, Carillion acquired Alfred McAlpine for £572m.

Stanton Iron Works Co Ltd

Originally started in 1846 when Chesterfield man, Benjamin Smith, and his son Josiah, brought three blast furnaces into production alongside the banks of the Nutbrook Canal. These original furnaces produced around 20 tons of pig iron – a basic type of iron – per day but the company experienced financial difficulties and there were several takeovers during the middle of the 19th century. The Stanton Ironworks Company Ltd and during the First World War Stanton produced large numbers of shell casings, while during the Second World War both shell and bomb casings, gun barrels, and concrete air-raid shelter components were made. With its experience in high quality concrete products, Stanton was also involved in the production of experimental concrete torpedo casings and made 873,500 bomb casings.

At its height, 7,000 people worked at Stanton, making it one of the biggest employers in the area. During the early 1980s Stanton became part of the French Pont-a-Mouson Group and later part of Saint Gobain. Production wound down during the 1990s and eventually stopped on May 24 2007, when the last pipe was made. Soon after the closure, Birmingham-based Company Spring started creating controversial plans to build 4,000 houses on the site. But in 2008, the company was forced to pull out of the deal after its residential property arm suffered badly in the credit crunch. Saint Gobain still wanted to continue with the plans to create the village on the brownfield site. After years of indecision, plans were finally being submitted for the approval of Erewash Borough Council.

Welsh Granite Company

The Penmaenmawr & Welsh Granite Co owned and operated a major granite quarry on the North Wales coast located between Conwy and Llanfairfechan. The quarry and its internal narrow gauge railway continued to thrive during the nineteenth century. Production at the quarry continues in 2006, though the railway was replaced by Lorries in the 1960s.

The Rhosesmor Sand & Gravel Company

Rhosesmor Sand and Gravel Company limited was registered on 1960.03.01. The company's status is listed as dissolved as of 31.10.1994.

Wm. Cooper & Sons, Ltd Mersey River Dredgers

Wm Coopers did have a berth at Warrington and they also had berths at Canning Dock, Liverpool where you could see the sand silos from the Mersey. The silos were situated on the north of Caning Dock. The only reference we can find is a company is in Thorpe Egham, Surrey and going strong.

Gresford Sand and Gravel Company, Ltd

The Marford Quarry, (northern site) was opened in 1927 to provide materials for the construction of the Mersey Tunnel, but quarrying ceased in 1971, when it was close by Tarmac Ltd. Twenty-six acres of the thirty-nine acre site were bought by the North Wales Wildlife Trust in 1990.

The quarry site was to be developed into a large housing site but the development didn't go ahead because it was found that the ground was moving. Planning for the sites restoration and the planting of trees, 9 possibly to stabilise the ground) was issued on October 28th 1980, the details of which were approved on March 12th 1982.

Indasco Bitumen Emulsion from the Royal Dutch Shell Group

Part of Royal Dutch Shell Group more commonly known for its petrol and oil refineries as Shell.

James B. Robertson & Co Limited

James Robertson of London, whose 'Marplax' finishes were selected for the initial coating of the tunnel interior, were equally successful in obtaining a contract for the supply of the final finishing surface of the walls of the fan chambers and switch-gear rooms, in which all exposed concrete surfaces have been treated with 'Stipplecrete', a glazed, stippled, impervious finish. No details can be found but a similar named company operate in Australia.

Francois Cementation Company Limited of Don Caster

Albert Francis was a Belgian man who had been working in England since his twenties. In 1910, he secured his first contract to sink a shaft in a colliery in Doncaster.

Despite Albert's attempts to use low pressure pumps and without chemicals his efforts failed and the contract was cancelled. In 1912, he succeeded in patenting his cementation process of grouting by pump pressure. The 1920s drew to a close with a record annual turnover of just under £4 million, delivering a record £404,000 profit before tax. In 1949, the Cementation Company (Canada) was incorporated. Work spread globally with contracts in the UK, Republic of Ireland, Rhodesia, Tasmania, India, Turkey, Pakistan, Italy, Spain, Portugal, the Gold Coast, Sierra Leone, Venezuela, France, Algeria, Tunisia and Morocco

Work for the National Coal Board was shrinking and general civil engineering work had become the mainstay of the company. At the same time, we were tentatively testing the building market with construction of a parabolic concrete sugar silo in Liverpool. This building was designed and constructed by Cementation, with a capacity of 100,000 tons of bulk sugar. When completed in 1956, this silo was the largest in the northern hemisphere and the second largest such structure in the world, measuring 165 metres by 51 metres with a height of 26 metres

After the boom years of the late eighties, the group, in common with the majority of British industrial companies, was reporting heavy losses. Between 1992 and 1993 Jardine Matheson, using their Hong Kong Land subsidiary, acquired 26% of the shares in Trafalgar House and attempted to turn the company around, but without success. In 1995, the losses amounted to £321 million, taking the accumulated losses in the first half of this decade to a figure approaching £1 billion. Rescue came to the ailing group in the form of Kvaerner ASA, a Norwegian shipbuilding and engineering group which had been unsuccessful in its bid for Amec in 1995. In April 1996, Kvaerner paid £904 million to acquire The Trafalgar Group. This new conglomerate had a turnover approaching $10 billion, with a workforce of fifty-eight thousand employees, part of the Skanska global strategy of being number one in your home market.

Skanska UK is part of Skanska, headquartered in Stockholm, one of the world's leading construction groups with expertise in construction, development of commercial and residential projects and public-private partnerships.

Wellington Haulage Liverpool to Dingle /Otterspool Waste Removal

Their history after the tunnels is unknown.

G & W Dodd

G & W Dodd of Birkenhead removed tunnel waste to Storeton Quarry Waste Removal but their history after the tunnels is unknown.

J. & A. Gardner And Co Ltd of Glasgow

Granite setts (Bonawe granite, which is quarried and trimmed on the shores of Loch Etive, on the north-west coast of Scotland). 28 April 1906 Quarrying of Ornamental and Building Stone, Limestone, Gypsum, Chalk and Slate last known returns. October 31 2008, company dissolved.

Perrin Hughes & Co Ltd of Bridge-Water Street Works in Liverpool

Their history after the tunnels is unknown.

There is a similar named company in Staffordshire (UK Architectural Antiques Ltd) but this is purely coincidental.

Bayley's Steel Works Ltd Sheffield

Much of the stainless steel was supplied by Brown one of the earliest firms to manufacture stainless steel.

Brown Bayley Steels was a steel-making company established in Sheffield, England in 1871, as Brown, Bayley & Dixon. They occupied a site on Leeds Road, which was later occupied by the Don Valley sports stadium. The firm was founded by George Brown, Nephew of 'John Brown' of the firm John Brown & Company. The firm manufactured Bessemer steel and railway tracks. Notable among its employees was Harry Brearley, the inventor of stainless steel.

Notable dates for the company are:

1937: Steel manufacturers. 'Hotspur' Stainless and Heat Resisting Steels. 'Twoscore' Stainless and Heat Resisting Steels. 'Weldanka' Stainless and Heat Resisting Steels.

1951: Brown Bayley Steels Ltd was nationalised under the Iron and Steel Act; became part of the Iron and Steel Corporation of Great Britain

1968: The Brown Bayley parent company, which owned Hoffman and Brown Bayley Steels, was taken over by the Industrial Reorganisation Corporation with a view to rationalisation of the sectors of roller bearings and alloy steels

On the night of December 15th 1940, a German air raid on Sheffield caused extensive damage to the steel works. In April of 2015, 2 x bombs were found on the now former site of the steel works.

Adams Brothers (Liverpool) Ltd

The treads of the staircases in the passages and on the landings were finished in granolithic, which was laid by Messrs.., who also covered the whole width of the concrete roadway in the tunnels with granolithic prior to the laying of the cast-iron setts.

Their history after the tunnels is unknown.

Ventilation Buildings and Equipment

It has already been mentioned that the whole of the steel required for the six ventilation stations was supplied by Redpath, Brown & Co Ltd of Manchester. But few amongst those who have seen the gaunt frameworks of steel girders, which rose into the sky, can conceive the immensity of the task which had been undertaken before the fabricated steel had been transported to the sites.

When the plans had been finally passed, it became necessary to construct models of every piece of steel that was to form part of the structure. Then the steel had to be fabricated to the required dimensions. Before it left the works, each section was marked to correspond with the working drawings supplied to the erecting staff. The speed with which the framework of these buildings was erected represents a triumph of organising ability comparable only with the efficiency with which each building was completed.

Trussed Steel Concrete Co Ltd

Smooth concrete finish of ventilation chambers and buildings.

Known to be still going strong in 1946 but no recent record of the company which may have been taken over by a larger company or group.

Redpath, Brown & Co Ltd

Redpath Brown and Co of Trafford Park, Manchester – steelwork in all six ventilation buildings

Notable dates for the company are:
- 1802: Company founded as Redpath and Brown, wholesale ironmongers, nail makers and iron merchants, by John Redpath and John Brown.
- 1880: Opened a department for iron constructional work such as beams and roof trusses.
- 1885: Introduced mild steel rolled joists.
- 1886: Made some steel joist compound girders, the first in Scotland.
- 1896: Public company. The company was registered on 15 October, to acquire the business of iron and steel merchants of the firm of the same name.
- 1896: Acquired site for the St. Andrew Steel Works in Edinburgh.
- 1923: Bolckow, Vaughan and Co acquired an 'important interest' in Redpath, Brown and Co makers of structural steel, with works in London, Manchester, Edinburgh, Glasgow, in order to secure an outlet for finished steel.
- 1929: Dorman, Long and Co acquired Bolckow, Vaughan and Co to create a major force in structural engineering
- 1937: Steel structural engineers and merchants.
- 1961: Iron and steel merchants and constructional engineers. 2,916 employees. Now believed to be part of TATA Steel

Ames & Finniss Brickworks

Suppliers of ventilation building bricks

Their history after the tunnels is unknown

Noted firms of contractors from Liverpool, Sheffield and London were awarded the contracts for the construction of the ventilation buildings.

W. Moss & Sons, Ltd
Built the North John Street station and their history after the tunnels is unknown

Messrs. Henry Boot & Sons Ltd
Henry Boot and Son of Sheffield, the New Quay and Taylor Street stations;

During the 1919–1939 inter-war period, Henry Boot built in excess of 80,000 homes, significantly more than any other contractor in the UK Pinewood Studios at Heatherden Hall, Iver, Buckinghamshire, was designed and constructed by Charles Boot as a Henry Boot development during 1936.

Based on the best US film studio practice and layout, the considerable work, including in-house design, was completed within a period of only twelve months. The year 1939 saw the commencement of a further spate of work for the Navy, Army and Air Force, including aerodromes, ordnance factory buildings and hospital camps (including what is now Haven Holidays Pwllheli) all of which were contracts where time was of the essence. Henry Boot also played a part, together with the country's other major top 20 contractors, in the monumental construction of the Mulberry and Gooseberry Harbour's for use in the invasion of Europe on D-Day, 6 June 1944

Expansion of Group activities continued, part of which was the acquisition of a foundry in Bingley, West Yorkshire in 1974 until its closure in 1983. Today Henry Boot is a fully listed company on the London Stock Exchange with a sound reputation for reliability and integrity in its specialist markets. The Company still retains its family links and the construction business specialises in serving the needs of commerce and industry. It is this flexibility and strength of performance that has kept Henry Boot at the forefront of its markets for over 125 years and which ensures that the Company is well placed to continue its impressive progress

John Mowlem & Co Ltd of London,
Founded by John Mowlem in 1822, the company was awarded a Royal Warrant in 1902 and went public on the London Stock Exchange in 1924.

A long-standing national contractor, Mowlem developed a network of regional contracting businesses including Rattee & Kett of Cambridge (bought in 1926); E. Thomas of the West Country (bought in 1965); Ernest Ireland of Bath; and the formation of a northern region in 1970. During the Second World War the company was one of the contractors engaged in building the Mulberry Harbour units.

Mowlem acquired SGB Scaffolding Group in 1986 along with Unit Construction in 1986, giving the firm a substantial presence in private housebuilding. The ensuing recession led to losses of over £180m between 1991 and 1993 and banking covenants came under pressure. The housing division was sold to Beazer in 1994. Mowlem was bought by Carillion in February 2006.

Electrical installation
Higgins & Griffiths Ltd
(George's Dock Building)
>High Court of Justice papers were filed in 1936.
>Their history after the tunnels is unknown.

John Hunter & Co Ltd
(North John Street and Taylor Street Buildings)
Their history after the tunnels is unknown.

Electric Power Installers Ltd
(New Quay and Sidney Street Buildings)
Their history after the tunnels is unknown.

Plumbing Work
Messrs. R. W. Houghton, Ltd, of Liverpool, were entrusted with contracts for the required at George's Dock, North John Street, New Quay, Sidney Street and Taylor Street ventilating stations.
Their history after the tunnels is unknown.

Double Doors to the Fan Chambers
The Birmingham Guild Ltd
Supplied special double doors to the fan chambers in all the ventilating stations. These were designed and fitted as an additional precaution against the transmission of sound and vibration. The outer doors were of hollow steel framing, whilst the inner doors took the form of hollow wood shutters, all being firmly clamped against vibration.
Their history after the tunnels is unknown.

Mr W. H. Law, of Messrs. W. M. Law & Son,
26, Exchange Street East, Liverpool, a member of the Chartered Surveyors' Institution and Member of the Quantity Surveyors' Committee of the Chartered Surveyors'.

All the electrical motors, transformers, switch and control gear, metres, protective relays, and connections inside the stations have been supplied by the Metropolitan-Vickers Electrical Company Ltd who have also acted as main contractors for the supply of reduction gears made by David Brown and Sons, Ltd, Huddersfield, and for hydraulic couplings, made by the Hydraulic Coupling & Engineering Company Ltd.

Of the fans, sixteen for the George's Dock, North John Street and Woodside ventilating stations were supplied by Walker Bros (Wigan) Ltd while fourteen fans for the New Quay, Sidney Street and Taylor Street have been supplied by the Sturtevant Engineering Company Ltd of London.

Vickers Electrical Company
In May 1917, control of the holding company was obtained jointly by the Metropolitan Carriage, Wagon and Finance Company, of Birmingham, chaired by Frank Dudley Docker, and Vickers Limited, of Barrow-in-Furness.

On March 15th 1919, Docker agreed terms with Vickers, for Vickers to purchase all the shares of the Metropolitan Carriage, Wagon and Finance Company for almost thirteen million pounds. On 8 September 1919, Vickers changed the name of the British Westinghouse Electrical and Manufacturing Company to Metropolitan Vickers Electrical Company.

The immediate post-war era was marked by low investment and continued labour unrest. Fortunes changed in 1926 with the formation of the Central Electricity Board which standardised electrical supply and led to a massive expansion of electrical distribution, installations, and appliance purchases. Sales shot up, and 1927 marked the company's best year to date. In 1928, Metrovick merged with the rival British Thomson-Houston (BTH), a company of similar size and basically the same product line up. Combined, they would be one of the few companies able to compete with Marconi or English Electric on an equal footing.

In fact, the merger was marked by poor communication and intense rivalry, and the two companies generally worked at cross purposes. The next year the combined company was purchased by the Associated Electrical Industries (AEI) holding group, who also owned Edison Swan (Ediswan); and Ferguson, Pailin & Co manufacturers of electrical switchgear in Openshaw, Manchester. Rivalry between Metrovick and BTH continued, and AEI was never able to exert effective control over the two competing subsidiary companies.

In 1947, a Metrovick G.1 Gatric gas turbine was fitted to the Motor Gun Boat MGB 2009, making it the world's first gas turbine powered naval vessel. The Bluebird K7 jet-propelled 3-point hydroplane in which Donald Campbell broke the 200-mph water speed barrier was powered with a Metropolitan-Vickers Beryl jet engine. Campbell succeeded on Ullswater on 23 July 1955, where he set a record of 202.15 mph (325.33 km/h), beating the previous record by some 24 mph (39 km/h) held by Stanley Sayres.

The rivalry between Metrovick and BTH was eventually ended in an unconvincing fashion when the AEI management eventually decided to rid themselves of both brands and be known as AEI universally, a change they made on 1 January 1960. This move was almost universally resented within both companies. Worse, the new brand name was utterly unknown to their customers, leading to a noticeable fall-off in sales and AEI's stock price. This allowed AEI to be purchased by General Electricity Company in 1967, which was later restructured into many other companies, including Marconi plc in 1999.

David Brown and Sons Ltd

Their history after the tunnels is unknown.

Sturtevant Engineering Co Ltd

Sturtevant Engineering became an independent company following WW1 as changed government policy forced a divestiture. The 1960s marked the beginning of a series of ownership and organisational changes starting with Drake & Scull's acquisition of them. Closure of Manchester, the industrial fan plant, came later in the decade and signalled their departure from that business.

In 1972, the company was merged with New Welbeck, another British vacuum company, to form Sturtevant Welbeck. In 1978, they reorganised into Sturtevant Engineering Holdings, Ltd (Crusher line) and Sturtevant Welbeck. Around this time, Sturtevant Welbeck was broken into separate divisions; Sturtevant Engineering & Manufacturing Ltd (portable vacuums) and Sturtevant Systems Ltd (fixed vacuum systems).

The crusher business was sold to Christy Hunt (now Christy Turner Ltd) in the early 1980s. Both vacuum divisions were sold in 1994 to the Scotland-based business development company, Clyde Blowers, which promptly closed the Sturtevant Brighton plant. The assets were transferred to Clyde Material Handling Ltd which organised a new Sturtevant vacuum division at South Yorkshire. Today Sturtevant is the leading industrial vacuum cleaner company in the Euro zone.

Toll Booths

The original Toll Booths on the then four entrances were protected from the vehicles by rubber-coveted concrete kerb units. For the surfacing of the outer gradients leading from the public motorways to the mouth of the tunnel, granite setts were used. The chosen type was Bonawe granite, which is quarried and trimmed on the shores of Loch Etive, on the north-west coast of Scotland, by J. & A. Gardner and Co Ltd of Glasgow. This granite is not only of the most durable quality when subjected to heavy wear, but in addition its surface is such that it is practically impossible to skid on it. In the region of a thousand tons of setts were required for the works.

Roneo Ltd

The interior steel linings and equipment of the toll booths have been supplied by Messrs. Roneo, Ltd.
Their history after the tunnels is unknown.

Higgins & Griffiths

The electrical equipment of the toll booths, the central lighting shafts and the parapet and wall lamps.
Their history after the tunnels is unknown.

Synchromatic Time Recording Co Ltd

The electrical clocks and calendars in the toll booths were supplied by the Synchromatic Time Recording Co Ltd.
Their history after the tunnels is unknown.

Walker Bros. (Wigan) Ltd

Weigh Bridges (road registers)
Notable dates for the company are:
- 1866: Company founded. See J. Scarisbrick Walker and Brothers
- 1923: Ventilating fan engine for The Severn Tunnel.
- 1949: was a member of the Walmsleys (Bury) Group
- 1961: Engineers and makers of mining and paper making machinery
  1971: Beloit Corporation of USA received 'A' shares in the company which could be converted in 1975 to give it control
- 1975: Almost all of the profit had been earned by the Italian subsidiary; the UK operation had suffered from the effects of inflation and production problems; the demand for paper making machinery was depressed
- 1975: Acquired by Beloit of USA, to produce a world-competitive group in paper making machinery, Beloit Walmsley.

Architectural Features

Liverpool Artificial Stone Co Ltd

The lamp standards at the New Quay and Birkenhead entrances are made of special centrifugally spun concrete with a white Portland stone finish.

Their history after the tunnels is unknown.

Penmaenmawr and Welsh Granite Company Limited

Took an interest in the Liverpool Artificial Stone Company in the 1930s.

Their history after the tunnels is unknown.

Thompson & Capstick

The Portland Stone portals are surmounted by immense carvings designed by, in association with the architect.

Born in Belfast. He was the son of Edmund T. Thompson (born c.1871 in Antrim), a stone carver based in Antrim. Edmund Charles worked as a clerk in a foundry when he was about 13, this may have been the iron dressing firm owned by his maternal grandfather. In the early 1920s, he moved to Liverpool and between 1924-30 attended evening classes at Liverpool School of Art. As he carved sculpture for the India Buildings, Water Street, Liverpool in 1923 (including a 'Neptune Crest'), it is likely he had learned the technique in his father's workshop and studied full or part-time at the art school in Belfast prior to this move.

From the late 1920s, Thompson built up a successful practice in partnership with George Capstick in Cypress Street, Liverpool. The major schemes they worked on include: sculpture and interior decorative details, Martin's Bank, Water Street (1927-32) working for Herbert Tyson Smith;(Birkenhead Cenotaph) sculpture for the Mersey Tunnel and George's Dock Ventilation and Control Station, Liverpool (1932-34); incised figures, musical motifs and gilded panels for the interior of Liverpool Philharmonic Hall, Hope Street (1936-39); exterior sculpture and interior features for the Regal Cinema, Rock Ferry, (1936) and the Halkin Street Cinema, Flint.

From the late 1930s, they obtained commissions outside Liverpool, among them the gilded lions and figures for the British Government Pavilion, Empire Exhibition, Glasgow 1938 and twelve carved panels for First Class Smoking Room, RMS Caronia, Cunard, (c.1945-46). After the war, Thompson was appointed Architectural Model Maker to the Ministry of Housing and Local Government, a post he held between c.1946-61. Amongst the project, he worked on was producing the models for the Arcon Prefabs.

In 1959, Thompson was awarded an MBE for his services to the Ministry of Housing and Local Government. He died at the General Hospital Croydon.

H. H. Martyn

The ornamental cast-iron work to the central lighting shafts, the toll booths, and the parapet and wall lamps was carried out by, whose work throughout displays the finest qualities of metal craftsmanship.

- 1930: Formation of subsidiary: the Gloster Coach and Sheet Metal Co Ltd which purchased the Carriage and Wheel Works Co of George St, Gloucester.
- 1933: Amalgamation with Maple and Co

- 1935: Ship decoration was a large part of the architectural decoration work including for the SS Queen Mary.
- 1936: A. W. Martyn, founder of Gloster Aircraft Co and chairman of V. P. Airscrews was appointed chairman of Aircraft Components Ltd
- 1936: A subsidiary company, Cheltenham Manufacturing Co was formed, based at the Sunningend Works, to undertake furniture manufacture.
- 1939: Had expanded into metal work; contracts in hand included for the SS Queen Elizabeth
- WWII the production of aircraft components was resumed at Sunningend by H. H. Martyn (Aircraft) Ltd including manufacture of parts for the De Havilland Mosquito
- 1967: Aluminium windows, doors and screens supplied for the Queen Elizabeth Hall and Purcell Room in London.
- 1971: Declining demand led to the end of trading

Messrs Mellowes Ltd of Sheffield and Messrs

Eventually, Mr Rowse recommended the utilisation of glass, and, after a thorough consideration of the possibilities of this material, in which. St Helens, played a considerable part, it was decided that this recommendation should be adopted, and contracts for the construction and erection of the dado were entered into with these firms.

Part of the international Pilkington Glass group.

Emergency Exit at Morpeth Dock

The treads of the staircases in the passages and on the landings, were finished in granolithic, which was laid by Adams Brothers (Liverpool) Ltd, who also covered the whole width of the concrete roadway in the tunnels with granolithic prior to the laying of the cast-iron setts.

Their history after the tunnels is unknown.

Bibliography

I would like to thank the following organisations and groups that have assisted me in the research for this book by allowing me access to their numerous records.
- Wonders of World Engineering magazine at http://www.wow-engineering.co.uk/

Underground Liverpool by Jim Moore, published by the Bluecoat Press in 1988. This is a fascinating account of all the Liverpool tunnels, including the Mole of Edge Hill, the many rail tunnels

Two booklets published by the old Mersey Tunnel Joints Committee. They are called *Mersey Tunnels – The Story of an Undertaking* and *Mersey Tunnels 2*.
- Wirral Archives for their numerous and vast documents availability
- Merseyside Museum Large Objects Store and documents library
- Liverpool Library for their numerous documents and Microfiche items
- Incredible Wirral – Liverpool Echo / Wirral News / Trinity Mirror
- History of Liverpool – Liverpool Echo / Wirral News / Trinity Mirror
- The Mersey Tunnel. The Story of an Undertaking – Mersey travel
- New Brighton Our Days Out – Liverpool Echo / Wirral News / Trinity Mirror
- River Mersey – Gateway to the World – Liverpool Echo / Wirral News / Trinity Mirror
- History of Liverpool – 800 Years Our City – John Belcham
- Story of the Mersey Tunnels (Officially Named Queensway) – Mersey travel
- Liverpool and Merseyside – Photographic Memory – Cliff Hayes
- Magic of the Mersey Tunnels (DVD) – Mersey Tunnels
- AA Motoring Trust – Tunnel Test 2000 – Automobile Association
- BBC Mersey Tunnels England – BBC Archives
- All about the New Mersey Tunnel (Pamphlet) – Mersey travel
- The Mersey Tunnel (Queensway) (Pamphlet) – Mersey travel
- The Story of the Mersey Tunnel Officially Names Queensway
- Mines and Quarries of North Wales – Jeremy Wilkinson
- 100 Years of cementation 1910 – 2010 – Skanska
- US Navy Diving Manual Rev 6 – US Navy (Decompression)
- National Archives
- Heritage England (Formerly English Heritage)
- Ordnance Survey
- Birkenhead Mersey Tunnel Approach Scheme, July 1969
- Stanton Iron Road and the Mersey Tunnel (Liverpool Museum)